6-23-11

Dear Bill +
    I hope ?
little book to be ? interest.
    May you always walk
on the narrow path until
the day you pass through
the narrow gate unto His
Eternal Love, Joy + Peace.
    God bless you and your
family always.
                In His Peace,
                Gm

Why You Were Born

The chronological story of your existence from
the beginning of creation to now

Gene J. Tischer

authorHOUSE®

AuthorHouse™
1663 Liberty Drive
Bloomington, IN 47403
www.authorhouse.com
Phone: 1-800-839-8640

First published by AuthorHouse   01/08/2011

ISBN: 978-1-4520-9000-9 (hc)
ISBN: 978-1-4520-9001-6 (sc)
ISBN: 978-1-4520-9002-3 (e)

Library of Congress Control Number: 2010915391

Printed in the United States of America

This book is printed on acid-free paper.

To my wife, Bobbie, my sons Jason and Tanner and my grandson Killian. Thank you for sharing your lives and love with me. May we continue to grow closer together each day as we strive to live out the purpose of our lives.

TABLE OF CONTENTS

PROLOGUE

This is my story of how, and why, you and I came to be. When our story is told in simple chronological order, it is an amazing narrative -- one that is far more imaginative than any fictional tale you could imagine. Each of us knows, from personal experience, that we have a sense of self – genuine self-awareness. We are able to reflect inward on our own personal existence. We also recognize that these personal reflections focus, from time to time, on basic questions common to all of us:

Why was I born?

Is there a specific purpose to my life?

Why do people experience pain and suffering; and why, so often, is there apparent inequity in the suffering?

Does God exist; and, if so, what is He like?

Do I continue to "exist" after I die; and, if so, what do I experience then?

These are very important questions; well worth asking and well worth the effort to try to answer to the very best of our abilities and insights. What follows are my answers to these questions. Admittedly, while many of the answers remain shrouded in mystery, I posit my most logical answers to these questions based upon years of inquiry, study and contemplation of data, insights and analyses gleaned from science, philosophy and theology.

I do believe that truth is one and that truth presented by a scientist or a philosopher or a theologian is simply the truth and that truths from these each of these various intellectual disciplines are essential to understanding why you and I exist. These three sources of truth provide the information required to see the whole picture of, and really understand the grandeur inherent in, and the real purpose for, our lives.

This is a good time, in human history, to write this book. Today, religion has no better helpmates than modern theoretical, particle and astro physicists and molecular and micro biologists. The physicists are positing and verifying literally incredible truths about our macro universe and the biologists are discovering equally unbelievable complexity in the micro universe of cellular and molecular life, proteins, enzymes, etc.

So, while this is not a science book, the first two chapters have a heavy emphasis on scientific content. This is done to demonstrate, in a very summary manner, that modern science is discovering wonders in our universe that are as incomprehensible as any faith-based proposition. Do not become bogged down in the scientific text; it is not intended to be studied as an end in itself here.

The point of the science content is to demonstrate two theses:

1. **Whether modern scientific investigations focus on the macro or micro universe, the footprints of an infinite intelligent designer appear increasingly obvious; and**

2. **Spiritual assertions, that seemed so hard to accept as rational propositions in the past, now appear more in line with our intuition and common sense than many of the amazing propositions and hypotheses our empirical scientists are asserting in their professional journals on a regular basis.**

So, now, sit back, relax and enjoy the most amazing tale you will ever read – the story of you. It is told in chronological order-- beginning at the very beginning of creation and continuing to today, tomorrow and forever. I have only one favor to ask. Please read this little book with an open mind and allow that it may just be the truth, until you have finished reading it. Then judge its reasonableness and probability for yourself.

CHAPTER I

BEFORE TIME BEGINS UNTIL JUST AFTER THE "BIG BANG"

Science without religion is lame, religion without science is blind.
Albert Einstein

Before the universe (and time) began, God the Father, God the Son and God the Holy Spirit is. He is, in His essence, Existence Itself. He is Who is. He is One Being, in three persons, and the relationship among these three persons is Love. We cannot understand this reality, for our experience is that each "person" -- a living being who possesses two faculties: intellect and free will -- is manifested as a single, individual (human) being. God is different. The mysterious reality of God is that He is one loving Essence existing as, and in, three distinct Persons.

Now, around 13.7 billion years ago, God creates our universe -- that is, He makes our entire universe out of nothing. That is what the word "create" means, to make something from nothing (as the Romans would say: "facere ex nihilo").

God, who is Love, decides to create our universe and us. Why He decides to do this is also a mystery. We can surmise that, as Love, He wants to share Himself with created beings who can freely love Him and be loved by Him. But no one really knows. We only know He does so decide because the universe and we do, in fact, exist. Since we are contingent beings --i.e., being that does not have to exist -- we must conclude that the chain of causation of creation, stops with a Being Who is Existence Itself. Without this logical conclusion, there is no rational basis to explain the fact that the universe really does exist.

God performs this creative act in a most imaginative fashion. He creates an infinitely small and infinitely dense amount of matter – a gathering of "stuff" so small and so dense that our intellects and imaginations totally fail us when we try to grasp its reality. It is so small and so dense that it is beyond the ability of our scientists to describe it in quantitative terms – a true "material" mystery.[1]

As part of the creative process at the start of the universe, the infinitely tiny and dense package explodes with such power that the low base sound generated by that burst of energy is still readily found, by specialized detection equipment, everywhere in space today. In fact, spectral and frequency analyses of this 13.7-billion-year-old residual energy wave continue to yield new discoveries about the formation of matter during the early years of the universe.

When this "big bang" occurs (the ultimate understatement on one hand, and the ultimate misnomer on the other (there is no "big" and there is no "bang")), the matter starts to expand. Theoretical physicists are constantly updating their knowledge of the early stages of the creation of our universe. But, as this is being written in the spring of 2010, these scientists are postulating a specific sequence of events in the initial creation phrase of the universe. At the very beginning a lot happens in a very short period time. In fractions of one second -- fractions so small that we would need 40 or 50 zeros to identify the sequence of events – the events described in the following table occur in the noted incomprehensible fractions of seconds.

General Physical Description	Detailed Description	Time Post Creation	Temperature
Creation		0	?
Quantum Gravity Era	Strings – other exotics	10^{-43} Second	10^{32} ^{0}C

---

1   I use the word "mystery" in both empirical and spiritual contexts. As stated in the Prologue, our scientists are now leading us to fantastic observations and conclusions about our physical world that are as mysterious and incredible as any ever proposed by a theologian.

Era of Inflation	Exponential expansion	10^{-35} Second	10^{28} ^{0}C
Electroweak phase transition	Electromagnetic & Weak Forces differentiate	10^{-11} Second	10 quadrillion ^{0}C
Quark-Gluon Medium	Liquid-like: elliptical flow and particle jet quenching	10^{-11} second to 10 microseconds	10 quadrillion to 2 trillion ^{0}C
Nucleosynthesis	Helium & other elements from Hydrogen	100 seconds	1 billion ^{0}C

It may well be that other dimensions and other universes arise at this time. The latest attempt to explain a unified theory of everything – the string theory – seems to postulate at least 9 or 10 dimensions and other universes outside our own. It is not the intent of this book to delve into the amazing, but tremendously complex, disciplines of theoretical and particle physics[2]. However, it is important to understand and acknowledge that, today, both scientists and men and women of faith look with awe at the creative process that brought about our universe and each make equally incredible statements about the process.

Scientists are limited by their discipline's command to not accept as true that which cannot be physically verified and duplicated through empirical experimentation. They are not allowed, as scientists, to ever make a "leap of faith." However, it does appear safe to say that their theoretical description of the universe's creation will never be verified or duplicated so many of their assertions will always remain, at least in part, technically rooted in the sphere of belief.

In any event, thanks to the "Era of Inflation," in the blink of an eye, following the "Bang," the universe is about the size of a baseball. Think

---

2   After you finish this little book, I strongly recommend you read *Beyond the Cosmos: What Recent Discoveries in Astrophysics Reveal About the Glory and Love of God* by Hugh Ross (PhD in Astronomy). The author explains how the many other dimensions that exist, as posited by modern physicists, make more rational and comprehensible the existence of God, His biblically described attributes, and His ability and capacity to interact with us in our 4-dimension universe.

about that for a moment. Everything we can see or detect, from the car on the street, to our home, to massive buildings and bridges, to large landscapes, to the earth, to our solar system, to the 100,000,000,000 stars in our Milky Way galaxy, to the 200,000,000,000 other galaxies -- each with billions of stars – all of this stuff, at one point in time, was contained, our scientists tell us, in a sphere the size of a baseball.

But, wait; that is not the most incredible aspect of the early stages of creation our modern scientists postulate, not by a long shot. Physicists also tell us that the "stuff" contained in that baseball not only includes all that we know as visible, detectable matter and energy; but it also includes a whole lot more dark matter and energy that exists but remains undetectable. Our scientists are certain that these dark components exist and that they comprise more than 95% of our universe. In fact, as far as we know today, the universe is comprised of the following:

- **Luminous matter**, such as stars, galaxies, nebulas – the objects we see when we gaze up at the night sky -- make up **.4% of the universe**. That number is correct – four tenths of one percent of the universe.
- **Ordinary matter**, which we can detect with various scientific instruments, is comprised of the many conventional atoms and sub-atomic particles and comprises **4% of the universe**.
- **Dark matter**, about which we know nothing since it is not detected directly by us and is not comprised of ordinary atoms, comprises **23% of the universe**. The scientific underpinnings for dark matter are:
  - Observations of the manners in which galaxies move – e.g., spiral band galaxies have outer stars moving so fast that gravity cannot hold them to the galaxy, but they remain there nevertheless due to the vast amount of unseen matter that must surround them and hold these outer stars in the formation; and
  - Theoretical models of the Big Bang indicate that this material cannot be composed of ordinary atoms – there are not enough of these atoms in existence to make up this dark matter.
- **Dark energy** makes up the remaining approximate **73% of the universe**. What is the evidence of this energy?

- ○ The universe is expanding in every direction at incredible speed and some force has to account for this fact; and
- ○ As stated above, we have detected the existence of primordial microwaves throughout the entire universe – the energy force left over after the Big Bang occurred.

Thus, we move quickly, riding on the mathematically fabricated wings of our theoretical scientists into a realm of the unbelievable, the incomprehensible; indeed, the unimaginable. When you have finished this story, *your* story, I sincerely doubt that any spiritual, faith-based assertion I make will appear more fantastic, impossible or unbelievable to you than this contemporary scientific statement: that, for a very brief moment in time, our entire universe -- luminous matter, non-luminous matter, dark matter and dark energy -- was the size of a baseball.

I wish I could more clearly communicate just how strange and wonderful this universe of ours is. Unfortunately, such communication requires knowledge of, and facility in, mathematics well beyond my capabilities; and, I assume, those of most of my readers. The fact is that the most accurate communication available to describe the magnificent strangeness of our physical universe is currently best comprehended by the few among us who understand advanced mathematics and relevant work products of scientists such as Albert Einstein and Werner Heisenberg. As Einstein writes: *[T]he theoretical physicist . . . demands the highest possible standard of rigorous precision in the descriptions of relations, such as only the use of mathematical language can give.*[3] and *[N]ature is the realization of the simplest conceivable mathematical ideas.*[4] Thus, I am left trying to convey the mysteries of our truly imaginative physical universe using linguistic metaphors as poor substitutes for mathematical expressions.

While it is true that even the language of math cannot totally capture the amazing reality that is our physical universe, it is also a fact that our physicists' magnificent and brilliant insights, expressed in mathematical terms, more accurately describe and communicate the unbelievable weirdness and counter-intuitive realities of our physical universe. After all, these scientists live in a world where:

---

3   Einstein, Albert. *Essays in Science.* New York: Barnes and Noble, 1934. p. 3. Print

4   Einstein, *op. cit.,* p. 17

- The speed of light is constant, which means the laws of physics appear the same to all observers, regardless of their states of motion (whether they are approaching each other or departing from each other);
- Time slows down as a body in motion speeds up;
- Journeys into and out of black holes are infinitely long (to the outside observer); and
- Time travel is possible (but interfering with an event crucial to your own existence creates a real cause/effects dilemma (perhaps solved by "closed time-like curves" whatever they are?)).

What is the point of this brief discourse on theoretical physics? First, it helps us agree that the creation of the universe is an amazing tale and interesting to think about just in scientific terms. Second, at times throughout this book I will make "spiritual assertions," such as God's role as Creator as stated above. These assertions require you to abandon any hope of really understanding or comprehending them. They are mysteries, just as the scientific assertion of the Big Bang event is, and will always remain, at least in significant part, a mystery. In both instances – one theologically generated and one scientifically based -- you are put in the position of making a reasoned-based leap of acceptance. Such leaps are the definition of faith.

This is an important point. Modern scientists ask us to take breathtaking leaps of faith equal to any required by a reasonable theological assertion. The demands made on our rational minds by modern science make it increasingly easy for us to consider the validity of our theologians' faith-based assertions as also possibly true. Truth is one, whether stated by a theologian or a physicist.

As science learns more about our wondrous physical universe, on both a macro and micro level, one has to admit that much of the content is beyond our common understanding. Scientific and theological mysteries abound and this will be true for your entire life. You just have to decide which scientific and theological assertions make sense to you; and, in both cases, you will end up making leaps of faith.

In any event, in our chronology, we now have a universe due to the Big Bang. This is an imaginative way to create a physical universe, to say the least. (It is also, in a very real way, "miraculous" in the sense that this gigantic explosion resulted, not in chaos, but in a level of precision that is beyond our imagination. See Appendix A.) Furthermore, note that

God's creative act is also indirect. His universe starts incredibly small; it expands and evolves over time. We all know that accomplishing a goal indirectly takes more intellect and imagination than just "doing it myself." So, too, is this true with God. His creative act is indirect in its effect, imaginative and intelligent – and each attribute to an infinite degree -- and containing a level of precision that is simply spellbinding.

CHAPTER II

FROM THE BEGINNING OF TIME
UNTIL LIFE FIRST APPEARS

Nature never deceives us; it is we who deceive ourselves.
Jean Jacques Rousseau

Once the universe is created, time starts, for time is merely the linear measurement of change and "linear," by its very nature, arises out of the concept of material space and matter. Space and time are both dimensional and relational. As time goes by, for some 9.2 billion years, the universe continues its expansion from the diameter of a baseball to the diameter of billions of light years. This is a large distance because light travels very fast indeed: 186,212 miles a second, 11,172,720 miles a minute, 670.36 million miles an hour, 112.6 billion miles a week, 488 billion miles a month and 5.86 trillion (with a "T") miles a year. Thus, 9.2 billion years after the Big Bang our universe has a radius of 53,877,894,819,840,000,000,000 (sextillion) miles.

At this point in time, about 4.5 billion years ago, in a relatively modest spiral galaxy – one among 200 billion galaxies -- far out on one of its spiral arms, a star is born. This collection of stellar matter proceeds to evolve into a medium-sized star, which comes to be known to us as our sun. The sun's gravitational field collects other spheres of matter in orbits around it, and slowly these materials grow to form our solar system's planets including, of course, the earth.

Early on, earth does not seem particularly favorable for any life form. For one billion years, from 4.5 to 3.5 billion B.C., the earth is forming in a sphere due to the gravitational tendency of matter to clump together. It is collecting material in its orbital path and being ceaselessly bombarded by these materials and incoming asteroids and comets. This period is referred to as the "Hadean Period" ("Hades", i.e., hell), when the Earth's surface is experiencing hot liquid plasma tsunamis and enormous volcanic eruptions all over its surface.

Needless to say, this environment is not favorable for life. However, over these one billion years, two fortuitous circumstances emerge that permit life to stake its claim. First, the earth cools to a reasonable temperature because it orbits the sun at the perfect distance for life as we know it. Second, a large "guardian" planet forms – Jupiter -- and it is located in a perfect position to allow earth to evolve in relative peace; protected, for the most part, from large outside bodies that could create terrestrial and atmospheric events that would interfere with the development of living entities.

So life begins. Now, we all use that word "life" in our everyday speech. But, what exactly is physical life? Why do we know that a tree is alive and a rock is not? Traditional scholastic philosophers define life as an entity with capacity for inherent change. This means the entity can make internal changes without the influence of an external actor or force. Aristotle, who would sometimes see the raw side of an issue, said life has two attributes: feeding and decomposing at death. Scientists define life as a being that has the capacity for auto-catalysis (the reaction product is itself the catalyst for that reaction), self-organization, spatial containment of functions and reproduction. Specifically, it has these characteristics (note that Aristotle got two of them right!):

- Storage of information -- contains genetic information;
- Breathing-fermentation -- transforms energy and matter taken from the environment into a form useful to its various parts;
- Internal stability – maintains necessary internal environment against external destructive forces;
- Boundaries – possesses flexible membranes that defines its parameter or the parameters of its constituent parts;
- Reproduction -- passes on its genetic information to offspring;
- Death and decomposition – genetics and external destructive forces.

Whichever approach you like best, each of these definitions has four basic characteristics:

- ○ Life "resides" in an entity – an individual, a unified organism with boundaries;
- ○ The entity executes inherent change by ingesting, breaking down and absorbing "energy packets";
- ○ It is able to pass on its genetic data to subsequent life; and
- ○ It dies and its material decomposes.

We do not know precisely what conditions are needed for life to come into being and we do not have any data on the first life forms. Earth is now very well suited to sustain life, but it has no special qualities or characteristics that explain the origin of life. Does God intervene directly to create the first living organism? There is no theological reason that requires Him to do so. He could have planted the seeds for life all those billions of years ago when He created the universe. However, my opinion, based on how new species appear as time goes by, is that God creates most life forms directly. He may use evolution to bring new species into existence from time to time – there is no theological reason why He could not use evolution to do so. But, as I discuss in detail below, data now clearly demonstrate that the evolution process is rarely used in the creation of a new species.

The earliest evidence of life is contained in fossil traces that date to 3.5 billion years ago. These earliest fossil remnants resemble a modern variety of aquatic and photosynthetic bacteria called cyanobacteria (blue-green algae). We owe a great deal to cyanobacteria for these microscopic entities were probably responsible for our oxygen atmosphere. And, while they are very tiny, they were the dominant life form on Earth for more than 2 billion years. Today there are a few places left where they still exist, for example in the Hamelin Pool in Western Australia.[5]

Scientists cannot tell us how life does begin. (Science, of course, cannot assert that God is the answer to this question.) What science does proffer, by way of an explanation, are two theories about this event. Each one presents significant problems with logic and known data. The first theory is simple and spectacularly unhelpful to us; the second is convoluted and replete with assertions and propositions for which there

---

5   This URL has interesting, detailed information on the amazing Hamlin Pool as of August 2010: http://www.australiascoralcoast.com/en/ Destinations/Shark_Bay_region/Pages/Hamelin_Pool_stromatolites. aspx).

is no empirical evidence. I include them here simply as an FYI and not as text you need to spend a lot of time analyzing.

1. Scientific theory #1 about how life began on earth. Life (or at least the chemical precursors of life) arose on other planets followed by a migration to various parts of the galaxy including Earth. This idea is know as panspermia. Recent support for this idea comes from Mars rocks found in the Antarctic that appear to contain fossilized organisms resembling bacteria. There is also some evidence from Martian probes suggesting the possibility of life existing either in the past or even currently on Mars. The problem with this theory, for our purposes, is that we still have the question of how that life began, whatever its original locus. Alien life, unless it is essential being (a definition for God), is only contingent being, like us, and not the ultimate explanation for how life began in the universe. This little book is about positing reasonable answers to your own life's ultimate questions and the panspermia theory gets you no further along in this quest.

2. Scientific scenario #2 about how life began earth. The second scientific hypothesis is "spontaneous generation." In this context, spontaneous generation means that inert matter developed into non-cellular macromolecular precursors that, in turn, eventually became living cells. This concept is expressed in various theories which, however expressed, simply posit a series of possible events and transformations, none of which is validated historically and most of which are not duplicable in the laboratory. Each of these theories is singularly unsatisfying and of no real help in our quest to understand why we are here on this earth. Therefore, I will not discuss in detail the various theories; rather, I strongly recommend, that if you ever want to study any such theories, you do a critical analysis to make sure that the following questions are answered:

A. Does the theory demonstrate reasonable ways that a complex molecule, such as DNA (deoxyribonucleic acid), could self organize? So far there is no credible scientific theory that explains how a structure of such complexity as DNA could arise from non-living matter. To get around this hurdle, some researchers theorize that RNA (ribonucleic acid) preceded DNA and this "RNA world" created the mechanisms needed to achieve DNA creation.

However, this sub-theory, if you will, also has gaping holes. RNA depends on proteins and enzymes to form. Therefore, the chicken and the egg question presents itself. So, even if we were to assume that investigators could prove that RNA was able to emerge in the absence of any enzyme causation, the theory would still have to demonstrate that the nucleotides could assemble into polymers and that the polymers could replicate without assistance from proteins since proteins would also not exist at this stage. Scientists have been struggling with this issue for many years and they are still unable to achieve the second step of replication -- copying a complementary strand to yield a duplicate of the first template -- without help from enzymes.

Moreover, there is another barrier to validating this theory. Scientists can induce copying of the original template only when they run experiments with nucleotides having a right-handed configuration. All nucleotides synthesized biologically today are right-handed. Yet, on the primitive earth, equal numbers of right-handed and left-handed nucleotides would have been present. When researchers put equal numbers of both kinds of nucleotides in their reaction mixtures, copying is inhibited.

B. Secondly, investigate to see if the theory explains, to any plausible degree, how aerobes (organisms that need free oxygen for sustenance) and eukaryotes (cells that have a nucleus) evolved. A theory that fails to detail the process that could have resulted in these astounding developments is of no use to us in our quest.

So where are we on the question of how life began? Scientists are no help to us. As we have seen, they posit panspermia or spontaneous generation. With regard to the latter theory, they can only tell us that, however RNA arose, its presence was probably the watershed event in the development of life. They assert that RNA very likely led to the synthesis of proteins, the formation of DNA and the emergence of a cell that became life's last common ancestor. They also admit that the precise events giving rise to the RNA world remain unknown. At this time, chemists, biochemists and molecular biologists admit that the details of how life originated remain a mystery whose secrets may never be revealed in a purely scientific investigation.

But there is an answer. The answer is that God creates life – after all, our common experience is that life only proceeds from life and God is essential and eternal life. Thus, about 3.5 billion years B.C., God creates life on planet Earth. Life first appears in very simple organisms; then, as time passes, God creates life in increasingly complex living organisms. But what about evolution as the explanation for the presence of increasingly complex and varied life forms?

For most of my life, I unquestioningly accepted the validity of the modern theory of evolution (Neo-Darwinism). I now know with certitude, based on the vast amount of factual data and advanced scientific discovery and analysis, that this theory makes no sense. In fact, it now seems a downright silly proposition. For today, it is obvious that only intelligent design accounts for the complexity and diversity of living species that have appeared on earth since life began.

The conclusion, based on the data we now possess, is that an infinite intelligent designer, known to many of us as God, is responsible for the millions of species the earth has experienced, for the most part, if not entirely, by His direct creative intervention. There may well be times when God acts through the indirect instrumentality or mechanism of evolution, but this does not happen very often.

What data and analysis drove me to this conclusion? First, and this is very important, my conclusion is not based on any theological principle. THERE IS SIMPLY NO VALID THEOLOGICAL OR PHILOSOPHICAL BARRIER OR OBJECTION TO THE THEORY OF EVOLUTION (except for the creation of human beings which materialistic evolution cannot explain for reasons stated below beginning on page 18). God is free to create new life varieties and species directly or indirectly through an evolutionary process. But the data now clearly demonstrate that He has chosen to intervene directly much more frequently than He acts indirectly through evolution.

If you fairly examine the evidence against the theory of evolution, it is overwhelming. The only scientific answer now possible to life's complexity and variety is intelligent design. Looking at all the evidence from a logical perspective, and not limited by the constraints of scientific methodology, it is apparent that God initially creates life and that virtually all new species come to be as a result of His direct creative act. As for the process of evolution, it may effectuate adaptive changes within a species or small group of organisms over short periods of time; but that is the extent of its role.

What are the data, analyses and logic that demonstrate, beyond any reasonable doubt, that the theory of evolution is totally inadequate to explain the knowledge we now possess? There are at least 5 categories of knowledge which demonstrate the errors in the theory. There are many other substantial deficiencies but these are sufficient to prove the point.

Let us understand what the theory posits. It theorizes that species gradually acquire significant and genetically transmittable biological improvements (i.e., mutations to their DNA) that enhance their chances of survival; and, through the application of the principle of natural selection, these species gradually evolve into entirely new species that are better equipped to deal with their environment. Since there are million of species that have lived, and do live, on earth, this transformation from one species into another has had to have happened millions of times. But the fact is that there are virtually no data to support such transformations. The theory way overreaches and is of no real value in any pursuit of understanding how life came to be as we find it today.

1. **The first failure of the theory arises from an examination of irreducibly complex mechanisms.**

In his book, *Darwin's Black Box*[6] microbiologist Michael Behe analyzes six irreducibly complex biological systems and demonstrates, with clear and convincing scientific data and analysis, how Darwin's theory fails in every way to explain how these systems could have evolved through the gradual step-by-step process that lies at the heart of the theory of evolution.

The author is well educated and experienced in this field. Professor Behe received his Ph.D. in biochemistry in 1978 from the University of Pennsylvania, where he did his dissertation work on sickle cell disease. He subsequently worked for four years at the National Institutes of Health on problems of DNA structure and joined the faculty at Lehigh University in Bethlehem, PA in 1985. He is currently a professor of biochemistry in the university's Department of Biological Sciences.

Professor Behe defines an irreducibly complex biological system as *a single system composed of several well-matched, interacting parts that contribute to the basic function, wherein the removal of any one of the parts causes the system to effectively*

---

6   Behe, Michael. *Darwin's Black Box.* New York: The Free Press, 1996. I highly recommend this book if you have any interest in the dizzying complexity of these amazing irreducibly complex bioprocesses. Professor Behe writes with excellent clarity and presents comprehensive details of his selected examples.

cease functioning.[7] The book's thesis is simple: relevant steps in biological processes occur ultimately at the molecular level; so, a satisfactory explanation of the creation of a complex biological phenomenon must include its molecular explanation. Any gradualism theory seeking to explain such a system's existence has to explain how irreducibly complex processes could be formed in a step-by-step process, how complex biochemical systems could be gradually produced. (Fossil records are of no help here -- they do not record changes at the molecular level.)

Darwin was well aware of this issue. He wrote: *If it could be demonstrated that any complex organ existed which could not possibly have been formed by **numerous, successive, slight modifications**, my theory would absolutely break down.*[8] (Emphasis added.) Professor Behe's book demonstrates Darwin's own identified theory killer in spades. He presents and analyzes six irreducibly complex biological systems that Darwin's theory is helpless to explain; yet they do exist. You will recognize them since they are operating, as you read this, in your own body. The six systems detailed in his book, with some notes about each, are as follows:

A. Immunization

The body's defense system is an irreducibly (and incredibly) complex process that reacts automatically to physical threats to our well being. It has to be able to distinguish the foreign invader from our own body. It does this by B cells in our bone marrow creating antibodies, each comprised of four chains of amino acids with binding sites that are designed to fit in and bind to a specific invading molecule. Since there are many different possible invaders, the body creates billions of anti-bodies for maximum protection. On average, it takes more than 100,000 anti-bodies to find one that will bind to a particular invader.

The system is so sophisticated that it can design a defense against even synthetically created protein invaders

7 Ibid. p. 39.
8 Darwin, Charles. *Origin of the Species.* New York: New York University Press, 6th ed, 1988. p. 154. Print

it has never experienced before. How is this possible? How can the body have a system that anticipates foreign bodies before those entities even come into existence? The answer is complicated, way beyond my understanding.   However I want to give you the scientific explanation in this lengthy (and exquisite) detail -- not because you should be expected to understand it; but because this creative aspect of our immune system demonstrates so well just how incredibly wonderful our bodies' irreducible biochemical systems have been fashioned by their intelligent designer.   Here is Mr. Behe's explanation (and remember, do not get bogged down -- just marvel):

The answer to the problem of antibody diversity had to await an astonishing discovery: a gene coding for a protein did not always have to be a continuous segment of DNA – it could be interrupted. If we compare a gene to a sentence, it was as if a protein code, 'The quick brown fox jumps over the lazy dog' could be altered (without destroying the protein) to read "The quick br*dkdjf bufjwkw nhru*own fox jumps over the la*pfeqmzda lfybnek sybagjufu* zy dog.' *The sensible DNA message was broken up by tracts of nonsense that somehow were not included in the protein. Further work showed that for most genes, corrections would be made – splicing out the nonsense – after an RNA copy is made of a DNA gene. Even with 'interrupted' DNA, an edited and corrected message in RNA could be used by the cells machinery to make the correct protein. Even more surprisingly, for antibody genes the DNA itself can also be spliced. In other words, DNA that is inherited can be altered. Amazing!*

. . . .

At conception there are a number of gene pieces in the fertilized cell that contribute to making antibodies. The genes are arranged into clusters that I will simply call cluster 1, cluster 2, and so forth. In humans there are approximately two hundred and fifty gene segments in cluster one: a ways down the DNA from cluster 1 are 10 gene segments that form cluster 2: Further down on the DNA road are a group of six segments that comprise cluster 3; and down a piece from that there are eight other gene segments that make up cluster 4. These are the players.

After the youngster grows a bit and sets his mind to getting born, one thing he wants to do is produce B cells. During the making of B cells, a funny thing happens: The DNA in the genome is rearranged, and some of it is thrown away. One segment from cluster 1 is picked out, apparently at random, and joined to one segment from cluster 2. The intervening DNA is cut out and discarded. Then a segment from cluster 3 is picked, again apparently at random, and joined to the cluster 1-2 segment.

The recombining of the segments is a little bit sloppy- not what you usually expect from a cell. Because of the sloppy procedure, the coding for a few amino acids (remember, the amino acids are the building blocks of proteins) can get added or lost period. Once the cluster 1-2-3 segment is put together, the DNA rearrangement is over. When it is time to make an antibody, the cell makes an RNA copy of the cluster 1-2-3 combination and adds to it an RNA copy of a segment from cluster 4. Now, finally, the regions that code for contiguous protein segments are themselves in a contiguous arrangement on the RNA.

How does this process explain antibody diversity? It turns out that portions of the segments from clusters 1, 2, and 3 form part of the binding site – the tips of the Y. Mixing and matching different segments from the three different clusters multiplies the number of binding sites with different shapes. For example, suppose that one segment from cluster 1 coded for a bump in the binding site, and another coded for a positive charge. And suppose that different segments from cluster 2 coded for an oily patch, a negative charge, and a deep depression, respectively. Picking one segment randomly from cluster 1 and cluster 2, you could have six possible combinations: a bump next to an oily patch, negative charge, or deep depression; or a positive charge next to an oily patch, negative charge, or deep depression. (This is essentially the same principle whereby pulling three numbers out of a hat explains the diversity of a state lottery; picking just *three numbers from 0 to 9 gives a total of one thousand possible combinations.) When making an antibody heavy chain, the cell can pick one of two hundred and fifty segments from cluster 1, one of ten from cluster 2, and one of six from cluster 3. Furthermore, the sloppiness during recombination 'jiggles' the segments (by crowding another amino acid into the chain, or leaving one out); this effect adds another factor of about 100 to the diversity [potential]. By mixing and matching DNA segments you get 250 x 10 x 6 x 100 which is about a million different combinations of heavy-chain sequences. Similar processes produce about ten thousand different light-chain combinations. Matching one light-chain gene to one heavy-chain gene at random in each cell gives a grand total of ten thousand times one million, or ten billion combinations! The huge number of different antibodies provides so many different binding sites that it's almost certain at least one of them will bind almost any molecule – even synthetic ones. And all of this diversity comes from a total of just about four hundred different gene segments.*[9]

As I stated above, understanding this process scientifically is not the point here. We just need to have a rudimentary grasp of its amazing complexity and ask ourselves how we could ever accept the assertion that this all came about through a **random mutation process**, **producing gradual results** and in less than 500 million years? That assertion defies common sense; yet, it is the essence of the theory of evolution.

B. Intra-cellular transport

A eukaryotic cell (one with a nucleus) is a very complex structure. It has many different compartments (organelles), each of which is sealed off from the rest of the cell by its own membrane. Some of the more notable components are:

- Nucleus (where DNA is);
- Mitochondria (produces energy);
- Endoplasmic reticulum (processes proteins);

---

9   Behe, *op. cit.* pp 127-129. (Footnotes omitted. Emphasis added.)

- Golgi apparatus (a way station for proteins being transported elsewhere);
- Lysosome (the garbage disposal);
- Secretary vesicles (store items before they are sent out of the cell);
- Peroxisome (helps metabolize fats).

In all, a cell has more than 20 different sections. Each of these areas has specific and precise interactions with other sections; and, some with the world outside the cell. These interactions are each, in themselves, irreducibly complex systems. Each time a delivery or pick-up is made there was a request made for the transport, the correct item for delivery or pick-up was identified, the correct component that can handle the item being transported was provided and then a delivery was made to the proper recipient all at the proper timing.

Each part is essential to accomplish the objective and remove any step and the transport process ceases to function. This irreducible complexity is demonstrated when an infant is diagnosed with I-cell disease. In this illness, enzymes needed to dispose of cellular garbage are misdirected away from the disposal compartment (the lysosome) and so the cell becomes bloated with old material that is not treated. It is a fatal disease caused by one misstep of one of the transport system components.

Our intra-cellular transport system is a wonder to behold. Furthermore, when one considers that these intra-cellular transports are occurring billions of times every second in your own body, we have to just stand back in awe at its precision and reliability. This system is like a miniature FedEx, only millions of times more complicated. Does anyone honestly believe that even a comparatively simple system like FedEx could just spontaneously arise with no planning or intelligent design? The answer, of course, is no.

On a tangential note, your body's 100 trillion (100,000,000,000,000,000) cells undergo 10 quadrillion (10,000,000,000,000,000) cell divisions during your lifetime. This is precision beyond our wildest imagination and defies any assertion that it all came about through some random gradual process.

C. **Paddling cells**

We also have complex cells with a special feature incorporated into them that allow movement. In some cases, such as sperm cells, the cell propels itself along. In other cases, such as cells lining our respiratory tract, the cell is stationary and an attached hair-like appendage, called a cilium, moves in synchrony with millions of its neighbors to keep harmful materials out of our lungs. The cilium is an amazing structure. It has many constituent proteins; e.g., tubulin, dynein, nexin and several other connector proteins. In fact, there are more than two hundred different proteins in a cilium. It has been, and remains, of great interest to many different scientific disciplines:

- Biochemists – its size and structure;
- Biophysicists – the dynamics of its power stroke;
- Molecular biologists – the expression of the many separate genes coding for its components; and
- Physicians – it occurs in some infectious microorganisms and it gets clogged in the genetic disease cystic fibrosis.

In spite of all this interest and research, there is no credible explanation, by anyone, on how this complex biological system could possibly have evolved step-by-step. It had to just appear, all at once.

D. **Formation of complex biological molecules**

Let us take one kind of molecule as an example -- the biosynthesis of AMP:

- Takes 13 precise steps;
- Utilizes 12 enzymes (one of the enzymes, IX, catalyzes two steps);
- Consumes:
  - 5 molecules of ATP for the energy to drive the chemical reactions;
  - 1 molecule of GTP;
  - 1 molecule of carbon dioxide;

- o 2 molecules of glutamine to donate nitrogen atoms at the proper times;
- o 2 formyl groups from THF at the proper times; and
- o 2 molecules of aspartic acid to donate nitrogen atoms at two other precise times.
- Requires, at 2 separate steps, remains of the aspartic acid molecule be cut off; and
- Requires, at 2 separate steps, parts of the growing molecule to be reacted with each other to close 2 ring shapes. (The functions and capabilities of proteins often depend on the physical shape of the entity.)

All of this complexity and precision is needed to just produce this one AMP molecule, one time.

E. Blood clotting

When you bleed, an amazingly complex and precise biological process kicks into action. It identifies the threat area; it sends help to that precise location and initiates a coagulation cascade. A net of material (fibrin -- created when the protein thrombin slices off, with great precision, specific pieces from the correct pairs of protein chains in fibrinogen) is formed that can trap escaping blood cells and stem the flow out of the body. The process also includes instructions, and this is important obviously, about when to shut down so the clotting mechanism does not migrate all over the body and shut off the blood flow to vital organs.

The odds that such a complex system could arise randomly are impossibly great for Darwin's gradual and random mutation premise to explain. Professor Behe calculates the odds thus:

Consider that animals with blood-clotting cascades have roughly 10,000 genes, each of which is divided into an average of three pieces. This gives a total of about 30,000 gene pieces. TPA [tissue plasminogen activator] has four different types of domains. By 'vigorous shuffling,' the odds of getting those four domains together is 30,000 to the fourth power, which is approximately one-tenth to the eighteenth power. Now if . . .

[a lottery] had odds of winning of one-tenth to the eighteenth power, and if a million people played the lottery each year, it would take an average of about a thousand billion years before anyone (not just a particular person) won the lottery. A thousand billion years is roughly a hundred times the current estimate of the age of the universe.[10]

F. Vision.

The eye is an amazing organ; again, comprised of multiple complex structures, each of which would have no purpose if any one of them were missing. There are many incredible, virtually instantaneous, reactions when we see. For example, when a photon hits the retina, the molecule 11-*cis*-retinal rearranges to *trans*-retinal. This action is the beginning of another lengthy (but incredibly fast), complex biological process that is irreducibly complex. Rather than go through another series of the bio-molecular steps in an irreducibly complex process – by now you have a basic appreciation for the complexity in a few of our body's irreducibly complex processes -- I want to present a different aspect of the miraculous process of our sight.

Vision begins with the photoisomerization (conversion of a photon of light into another form of energy) in a molecule. When the 11-cis-retinal molecule absorbs a photon, it isomerizes from the 11-cis state to the all-trans state. Now, suppose I were to ask you how long it takes for the molecule 11-*cis*-retinal to rearrange to *trans*-retinal? The answer is simply amazing -- just a few picoseconds. How fast is that? A picosecond is one trillionth, or one millionth of one millionth, of a second (0.000 000 000 001 seconds). It is about as long as it takes light to travel the width of a single human hair! As you receive the photons that are bringing to your eye the words on this page, your eye is performing chemical reactions at virtually the speed of light. Think about that for a minute (or 60 trillion picoseconds!). How wondrous is that?

In any event, and in conclusion, Darwin recognized

10 Behe, *op. cit.* pp. 93-94. (Footnotes omitted.)

that natural selection, the engine of his theory, only works if there is a mutation that is useful immediately, not sometime in the future. It is beyond dispute that irreducibly complex systems have separate constituent parts that are only useful when they are all present. The clusters of proteins in these complex systems have to be inserted all at once and this is not possible under his theory. Thus, Darwin's theory fails to explain these irreducibly complex systems.

As one final note about our amazing bodies, consider:

- The DNA of a single cell contains so much information that if it were represented in printed words, simply listing the first letter of each base would require more than 1.5 million (1.5×10^6) pages of text;

- If laid end to end, the DNA in a single human cell measures 3⅓ feet or 1 meter;

- If we could uncoil the entire DNA within a human adult's 100 trillion (10^{14}) cells, it would extend more than 63 billion (6.3×10^{10}) miles. This distance reaches from the earth to the sun and back 340 times; and

- Every one of our cells, except for our sex and blood cells, make 2,000 new proteins a second from hundreds of amino acids. The protein formation process is so complex that a super computer, programmed with the rules for making proteins, would take 10^{127} years to generate a final form for a single protein from just 100 amino acids.

Does it really seem possible that this degree of incredibly precise, yet spell-binding complexity can be subject to a **gradual, random-mutation selection** evolutionary process to explain the vastly increased complexity and diversity of life over the past 530 million years? This is a more unreasonable position than any theological assertion ever uttered.

2. **The unattainable mathematical odds against one species evolving into another.**

G. Ledyard Stebbins, a leading botanist and geneticist and one of the architects of the neo-Darwinian theory, estimates that that it requires 500 successive steps, each one beneficial, to transform one species into another. Ignoring the fact that all mutations observed to date are deleterious to the organism (please re-read this fact again), the likelihood of such a succession is miniscule to say the least.

But wait, there is more. Biophysicist and author Lee Spetner used experiments to determine that the odds a mutation will occur in a nucleotide transcription are 1 per 1,000,000,000 transcriptions.[11] Furthermore, the odds against one mutation appearing and surviving through the entire 500 successive step process are 1 in 300,000. This means, in the end, that the odds against one species evolving into another per Darwin's theory is 1/ followed by 2,700 zeros. Appendix B has a spreadsheet containing these odds. Look at it – is it reasonable to believe that these odds were satisfied millions and millions of times within the past 500 million years?

3. **The "spontaneous appearance" of many species with no fossil antecedents.**

An excellent example of these data is found in the Cambrian-era fossil records at 530 million B.C. Before this time, there are very few fossils of multicellular creatures. Then, with the passage of only 10 million years (the blink of an eye in geological and evolutionary timeframes), an explosion of multicellular life forms appears. Not only does the fossil record graphically present this "biological big bang" (as some have characterized it) in terms of quantity, but it also demonstrates that it was a qualitative explosion as well, in that these new multicellular living organisms are created with many varying parts and different, much more complex, body structures from any previous life forms.

How do evolutionists explain these data? They don't. For example, the website for Berkeley University, a organization not known for its leanings towards intelligent design, states: *Around 530 million years ago, a wide variety of animals burst onto the*

---

11   Spetner, Lee. *Not by Chance.* New York: Judaica Press, 1997: Print.

evolutionary scene in an event known as the Cambrian explosion. In perhaps as few as 10 million years, marine animals evolved most of the basic body forms that we observe in modern groups.[12]

You can see immediately how devastating this assertion is to Darwin's theory of gradual mutation selection. The website authors must have also realized this because they added an "interpretation" of the statement quoted above in an attempt to modify its impact and reconcile it with Darwin's theory: *The term 'explosion' may be a bit of a misnomer. Cambrian life did not evolve in the blink of an eye. The Cambrian was preceded by many millions of years of evolution, and many of the animal phyla actually diverged during the Precambrian.*

The problem is that there are also "many of the phyla" that have no identifiable fossil antecedents, in the Precambrian or any other period, and even the limited antecedents that do exist are only problematically and tenuously connected to a limited number of the new life forms that emerge. As the website further states: *Animal fossils from before the Cambrian have been found. Roughly, 575 million years ago, a strange group of animals known as Ediacarans lived in the oceans. Although, <u>we don't know much about</u> the Ediacarans, <u>the group may have included ancestors</u> of the lineages that we identify from the Cambrian explosion.* [Emphasis added.]

This statement is fanciful thinking. There is no scientific data to support its assertion about relevant antecedent fossils. That is why the assertion is so qualified as to be useless as a scientific statement.

Another notable example of the spontaneous appearances of thousands of new life forms, with no established fossil antecedents, occurred much closer to our own time -- 160 million B.C., with the Jurassic disaster. At this time dinosaurs, and <u>more than 80 percent of all species living at the time, met their extinction</u>. The current theory is that a meteorite struck the earth (near the Yucatan peninsula) and so changed the climate that few species were able to survive. In any event,

---

12 The Website URL, as of July 2010 is: http://evolution.berkeley.edu/ evosite/evo101/VIIB1cCambrian.shtml.

thousands of new species quickly appear on the scene with no ancestors to explain their presence.

The theory of evolution is helpless to explain these facts. In Darwin's own words: *As natural selection acts solely by accumulating slight, successive, favorable variations, it can produce no great or sudden modifications; it can act only by short and slow steps.*[13]

4. **No fossil record shows evolution from one species to another**

The known fossil record fails to document a single example of phyletic [gradual] evolution accomplishing a major morphologic [pertaining to the structure of an organism] transition and hence offers no evidence that the gradualistic school can be valid.[14] (Emphasis added.)

Scientists have now collected and catalogued more than 100 million fossils representing more than 250,000 species. On average, a species persists, fundamentally unchanged, for more than one million years before it ceases to be (often through catastrophic events). This well-established fact -- that species insist on remaining for long periods of time in their original configuration – presents another significant Darwin theory conundrum perhaps best summarized by University of Georgia geneticist John McDonald:

The results of the last 20 years of research on the genetic basis of adaptation has led us to a great Darwinian paradox. Those [genes] that are obviously variable within natural populations do not seem to lie at the basis of many major adaptive changes, while those [genes] that seemingly do constitute the foundation of many, if not most, major adaptive changes apparently are not variable within natural populations.[15]

---

13 Darwin, *op. cit.,* p. 309.

14 Stanley, S. *Macroevolution.* San Francisco: W. H. Freeman & Co., 1979. p. 39. Print.

15 McDonald, John. *"The Molecular Basis of Adaptation."* Annual Review of Ecology and Systematics 14, 93, 1983: Print. For much more detail on points 2-4 in this section read section 1 in James Hogan's excellent, thought-provoking book *Kicking the Sacred Cow.* New York: Simon & Shuster, 2004: Print

5. Human capacity to abstract

The fifth failure of Neo-Darwinism is derived from natural philosophy and logic. We human beings have an amazing capacity, one not found anywhere else in nature. We can think in abstract thoughts. This means we can "cut away" material notes or characteristics and see the essence of a material object. This ability to abstract is carried out at three levels:

A. We can form a "universal idea" in our mind. Thus, we can recognize the concept of "treeness" in a wide variation of tree shapes, types and textures – in the more than 100,000 species of trees that live in our world today. The tree may be a pine, maple, banyan, white spruce, fig or some other type; to us each is recognized as a tree. Our intellect is somehow able to abstract from the accidental particulars of each one (e.g., leaf shape and color, bark texture, size, shape, etc.) and we intuitively know that each one is a tree, an organism that is appropriately called by the same general name as all other trees. Our mind identifies the essence, the "treeness" if you will, in each tree and logic allows us to assign the same word to every member of this amazingly extensive category of plant life we call "tree."

B. The second level of abstraction is mathematics. At this level the material universe is represented by numbers and symbols that relate to each other as words relate to each other in text. There is a further reduction in the presence of matter's accidental characteristics and the mind probes to form a deeper understanding of the essence of matter. Thus, at the very lowest level of mathematical content, we say 2+2=4 without indicating any particular reference to what exactly is counted. At the very highest level, mathematicians fill notebooks with symbols and numbers extrapolating, for example, from the elegant simplicity of $E=MC^2$. Aristotle observed that a number has two meanings: that which is counted (or countable) and that capacity by which we count. And "that capacity" is a wonder of our intellectual faculty.

Einstein described this second level of abstraction thus: *The theoretical scientist is compelled in an increasing degree to be guided by purely mathematic, formal considerations in his search for a theory, because the physical experience of the experimenter cannot lift him into the regions of highest abstraction.*[16] (Emphasis added.)

C. The last level of abstraction is where thoughts are created that have no material notes or characteristics at all and includes our amazing power of self-awareness. Thus, we write poems about love, sing hymns to God, warn one another about evil and try to uphold virtues such as loyalty and generosity. These thoughts are independent of any material universe; in fact, they often are created in meditation where the world's presence is intentionally reduced as far as possible. These thoughts are capable of being purely spiritual and immaterial – totally devoid of any materiality content or context.

What does our capacity to abstract to the third degree mean for the theory of evolution? It means we humans have been caused by a cause that can also abstract to the third degree and Who must therefore possess immaterial (i.e., spiritual) capacities. This means He is an intelligent being Who knows and chooses. Here is the logic behind this assertion:

We are contingent beings; that is, we do not have to exist. In fact, as we have seen, there was a time when the universe itself did not exist. That means we are effects, objects caused by something or someone else. We did not cause ourselves. But there is a scientific principle that states that no effect can be greater than its cause. Therefore, whatever caused us must have the capacity for immaterial abstraction that can be passed onto the effects, the objects, of his causation (i.e., us).

Now Darwinism stands for the proposition that all life forms evolved from previous life forms. But the capacity to abstract to the third degree cannot be caused by life forms antecedent to us because none of them had this superior capacity and that would render the effect greater than its cause. Therefore, Darwin's theory fails to explain what caused our ability to abstract to the third degree, where content is immaterial.

Conclusion. Neo-Darwinism fails to explain our data and knowledge of the complexity and diversity of living things. But the

---

16 Einstein, *op. cit.*, p. 69

theory lives on (no pun intended) as an embarrassingly obvious ideology due to the stubborn reluctance of evolutionary biologists to just bite the bullet and state what is now clearly demonstrable – that living species' complexity and diversity is a result of intelligent design and not of gradualistic, random evolutionary forces. Our biologists should have the integrity to admit this and then add that they, as scientists, cannot provide any scientific information or data about the designer. This would be the honest position; but one they have so far failed to assert.

As an interesting aside, contrast the openness and honesty of our best physicists, who had vested much time and effort analyzing Einstein's general theory of gravity, with the dishonesty and politicization of our evolutionary biologists who have a commensurate vested interest in supporting Darwin's theory. As it became apparent that Einstein's theory does not explain or account for data derived from quantum mechanics, the precise measurements of the moon's orbit, etc., our physicists accepted this reality, acknowledged the weaknesses in Einstein's theory, and then set about the arduous process of formulating a more accurate theory for gravity that fits all the known data.

However, our evolutionary biologists refuse to make the same difficult decision to discard their equally highly regarded Neo-Darwinism postulate, despite overwhelming data uncovered in the last several years that renders this theory fundamentally erroneous – indeed, dare I say it, even moronic. It is ironic, indeed, that these so-called scientists are acting more like the Catholic church leaders in the 16th century, who rejected Copernicus's and Galileo's data due to a priori conceptions they simply refused to discard. Such blind denial is, unfortunately, the state of evolutionary biology today.

Of course, it may also be that because Einstein himself was open to the idea of intelligent design behind our universe, his fellow physicists were more open to follow the truth wherever it might lead than is the case with Neo-Darwinists. In his *Essays in Science* he writes:

Certain it is that a conviction, akin to a religious feeling, of the rationality or intelligibility of the world lies behind all scientific work of a higher order. This firm belief, a belief bound up with the deep feeling, in a superior mind that reveals itself in the world of experience, represents my conception of God.[17]

---

17 Einstein, *op. cit.,* p. 11.

One further note: Sir Isaac Newton, physicist and mathematician and Werner Heisenberg, the creator of quantum mechanics, also concluded that the universe was the product of an intellect with no limitations. These three men, arguably the three greatest scientists in history, each arrived at the exact same doorstep – that of an intelligent designer.

CHAPTER III

FROM THE CREATION OF LIFE UNTIL THE ARRIVAL OF JESUS

In any event, life begins and, as time passes, various species appear and life's diversity and complexity increases, sometimes exponentially. In sum, this is the time sequence for major developments leading up to human beings and then to the one perfect human being, Jesus Christ.

3.5 billion B.C. – Life begins in simple organisms. They reproduce right from the start. This is note worthy because reproduction is a purposeful act – a species need not reproduce except for the reality of its own death.

1.5 billion B.C. -- The first eukaryotic organisms (cells with a true nucleus).

530 million B.C. – "Cambrian explosion" event. We have briefly discussed what transpired at this time in the previous chapter. One of the major developments is the arrival of sex. New species arise like crazy and many of them are essentially different from any antecedent species that existed before.

200 million B.C. – Pangaea, the supercontinent that contains virtually all of the landmass of the earth, breaks up. Fifty percent of all plant and animal species disappear.

160 million B.C.-- The Jurassic disaster results in more than 80 percent of all species living at the time becoming extinct.

7 million B.C. – The first animals that crudely resemble us human beings.

500,000 B.C. -- Homo sapiens. They look more like us; but the brain is significantly smaller than ours. They use crude stone tools and make use of fire for the first time.

130,000 B.C. -- Homo neanderthalensis. They are large, bulky, human-appearing animals and they provide the first evidence of caring for the sick and burying the dead.

150,000 to 50,000 B.C. – Homo sapiens sapiens. The first modern human being arrives. (There is a wide range in the timeframe because it is derived from imprecise probability calculations of mutation rates of human genetic material.) At this time, a momentous change takes place in mental abilities and capabilities.

Some scientists characterize the appearance of the first modern human beings as the "Great Leap Forward." From their empirical scientific perspective, two issues are unresolved:

1. **What was the triggering cause?** Scientists theorize that one of two physical changes in the body may have resulted in this enhanced mental acuity:
    a. The perfection of the voice box allowed more complex and sophisticated sounds which resulted in the desire for better communication and this caused the brain to respond by developing more complex neural pathways so language could be mastered; or
    b. There was a change in the brain's organization so it was able to function to a much more advanced degree than was possible before.

This is a far as modern science can go in its speculation on what caused the human brain to make an incredible qualitative leap in its ability to be self-aware, and to know, judge, evaluate, contemplate, etc

2. **Where did it take place?**

a. One scientific theory is that it occurred in one location and then spread. This is supported in three ways.
   i. Modern-looking skulls dating from around 100,000 years ago have been found in Africa;
   ii. Some molecular studies of mitochondrial DNA argue that we all had one common female ancestor; and
   iii. When we investigate how the Cro-Magnons displaced the Neanderthals after the latter occupied Europe for thousands of years, we have a concrete example of how the superior modern human beings would have replaced all previous humanoids around the world.
b. The other scientific theory about location is that the appearance of modern human beings occurred simultaneously at various locations around the world. This theory is based on the fact that fossilized human skulls in China and Indonesia are considered to have features of modern Chinese and Australians respectively.

Science has no satisfactory answer to account for the "Great Leap Forward." However, there is an answer. As God creates increasingly complex living species, He reaches the point when the neural complexity of the brain is sufficient to carry out the very high-level functions of intelligence and free will. He then directly creates the first human soul, a component that can take full advantage of this complex, neurologically enhanced brain; and, acting in concert with the brain, the soul and body combination can now perform intelligent and willful actions. Thus is created the first human being, a man whom we shall call, for convenience sake, "Adam." About the same time, God also creates a female human being whom we shall call "Eve."

Adam and Eve are a new type of living organism – they are modern human beings, they are "persons." I briefly defined "person" at the beginning of Chapter One when writing about the Trinity. Adam and Eve are persons because each one is a living being who has the two requisite incredible powers, the two amazing capacities – intelligence, i.e., the power to know to the third level of abstraction; and free will, i.e., the power to choose. These two powers are immaterial, spiritual, powers that define personhood; and an immaterial, spiritual agent, the human soul, makes them possible.

Thus, Adam and Eve have something essential in common with God. God and Adam and Eve each have the faculties of intellect and free will (although God's are infinite and perfect). In this sense, it is accurate to say that they are created in God's image and likeness. They are all "persons."

Physically, too, Adam and Eve are qualitatively different from any other living organism that has preceded them because they each have a spiritual soul that infuses a body that has a brain that is the most amazing material organ in creation, and this on several levels:

- Physically, it weighs about 2.8 pounds (44.8 ounces). It is 77% cerebral cortex, where our higher functions are performed; that makes it three times larger in humans than in Chimpanzees (another fact that is hard for evolutionists to explain since we share about 95% of our genetic material with chimps). Even more amazing, it is comprised of 78% water (35 ounces), 12% fats, 8% proteins and 2% salt and carbohydrates. We human beings think with an organ whose non-water component parts weigh less than 10 ounces. How incredible is that? (It would appear that God did not leave Himself much brain material with which to work; but it is obviously enough, for you and I are communicating with our own 10 ounces of non-water brain material right now!)
- It allows Adam and Eve and us to cogitate at the third level of abstraction.
- It provides Adam and Eve and us with the instrument the soul needs for self-awareness.

What does that mean? Adam and Eve, and each one of us, is conscious that we exist and that we are separate and distinct from the rest of creation. We have first-person, subjective experiences every day.

This personal experience of self-awareness is the neural mystery for science bar none. Science cannot discover how the brain creates self-knowledge from studying the physical brain. But, for science, there is nothing else to analyze, to study.

In this discussion of self-awareness, psychiatry and empirical psychology offer no assistance to our quest to understand our own story. These disciplines are interested in the process, the results, the variations, etc. of our mental activities. We seek to understand our very selves, our essence, including the phenomenon of self-awareness itself, its content; how it is that each one of us, beginning with Adam, is aware of himself or herself as a unique, individual, being. Wherein comes that capacity?

Self-knowledge is apparent to each one of us in at least two aspects:

- Each one of us experiences the fact of our own existence as a reality in our daily lives; and
- Each one of us recognizes our own mental dispositions and habits as those of an individual human being.

The great psychologist William James acknowledges each person's ability to perform self-awareness while admitting that science is not capable of empirically demonstrating it exists or how we accomplish this amazing feat. He writes: *All people unhesitatingly believe that they feel themselves thinking and that they distinguish the mental state as an inward activity or passion, from all the objects with which it may cognitively deal. I regard this belief as the most fundamental of all postulates of Psychology, and shall discard all curious inquiries about its certainty as too metaphysical.*[18] (Emphasis added.)

That leaves establishing the certainty and understanding of self-awareness to theologians and philosophers who fear not to tread where metaphysics (literally "after physics" – i.e., knowledge beyond that derived from the empirical sciences) leads them. And they have not ignored the issue. The perplexing issue of self-awareness has been studied over the centuries by many of the best minds in history, culminating with St. Thomas Aquinas, who created the most sophisticated, analytical presentation of self-awareness as a topic, recognized and analyzed as such, that has ever been done.

The answer to the question -- What is self-awareness, self-knowledge? -- is well stated by professor Richard Lambert, in his book entitled *Self Knowledge in Thomas Aquinas*:

There are two kinds of particular cognition – actual and habitual. People have actual knowledge (actualem cognitionem) of their possession of a soul because they perceive the soul through its acts; someone knows that he lives, and thus that he has a soul, because he notices that he senses and

---

18 James, William. *The Principles of Psychology.* New York: Dover, 1950. c. VII, p. 185. Print. [Emphasis added.]

thinks. Cognition of objects remains primary, as we know a thing before we know that we know the thing. On the other hand, the soul has habitual knowledge (habitualem cognitionem) of itself through its essence: the soul can reach actual cognition of itself simply because its essence is present to it.[19]

This means that self-awareness is an immaterial, if you will, spiritual, act; and, as such, it is an action that can only be achieved by an immaterial, spiritual capacity or agent within us -- our soul. Thus, the reason you and I have self-awareness is because we have a soul that, through its spiritual power, knows itself and since it infuses our body that self-knowledge is present to our consciousness so we understand it and know that we are self-aware.

Why does God create Adam and Eve with these wonderful new capabilities? The ultimate answer is a mystery; nobody really knows the mind of God. But we can reason to a logical explanation that may provide a somewhat useful response to this question.

God is love and love is the touchstone of His created universe. So, God's universe is created to include created beings who can truly love Him, because they do so freely, and be loved by Him. To be free to choose, these beings must possess intelligence and free will; i.e. be persons who have the ability to know and the capacity to freely accept or reject the known, including God Himself.

Therefore, out of love, God creates contingent beings (i.e., beings who do not have to exist) who have the power to freely love or to freely reject Him and His love. He does not overwhelm Adam and Eve by making Himself so obvious and so dominating that they have no choice but to know He is and to do His will. Instead, He interacts with them in an indirect manner, so if they want to love Him and do His will they will have to make an effort to learn about Him and then choose, freely, to love Him.

So, Adam and Eve are created with the faculties to love and obey or to choose their own will over God's. Unfortunately, it comes to pass that Adam and Eve make a bad choice with their free will; and, as a consequence, experience a diminishing, a blunting, a weakening of their

---

19 Lambert, Richard Thomas. *Self Knowledge in Thomas Aquinas.*
Bloomington: AuthorHouse, 2007. p. 36. Print. This is a work
of first-class scholarship -- articulate, logical in presentation and
comprehensive.

intellects and free wills and a distancing of themselves from the reality of God Himself.

How do we know this happened? Well, first, just watch the evening news. War, strife, crimes, injuries, etc., predominate. Also, each one of us knows well, from our own common sense and experience, the internal war that rages inside ourselves. We often have to fight to know the good and to do the good. We struggle with bad impulses and motivations. We deliberately hurt others, and then feel badly that we did not behave better; then we go out, and do it again.

What bad choice do Adam and Eve make and why do we have to suffer for it? No one knows the exact bad choice they make. In a theological context, the story in the Bible's Book of Genesis says that Adam and Eve eat the fruit of the tree of knowledge of good and evil in order to be like gods themselves. This imagery is an ingenious way to present the essence of the truth: Adam and Eve did something out of pride that set them against God.

When Adam and Eve go against God's directive, whatever it may have been, God, being a creator perfect in all respects and true to Himself, has to accept their choice. Unfortunately, that also means that God, since He is perfect in all His attributes, including that of justice, has to allow the deleterious consequences of their decision to occur.

This is an important point. God is a God of love and mercy, true. But He is also infinitely just, and justice – the rendering to each person what is his or her due – requires consequences for bad free-will choices. There are four substantial adverse consequences that follow from Adam and Eve's bad choice, each of which is transmitted down to every one of their descendants, including you:

1. They impair their faculties of intellect and free will. They, and we, have weakened faculties and encounter more difficulty knowing the good with our intellect and choosing the good with our free will than God originally intended. This accounts for how difficult it is for us to deal with the internal strife each one of us experiences every day between our good and bad impulses. We are not fully unified persons any more. We do not have the unanimity between intellect and free will that Adam and Eve enjoyed initially and these two faculties are now no longer in perfect harmony and sync.

As a result, we often do that which we know we should not; and we often do not do that which we know we should. It is one thing to know the good; it is an entirely different matter to always do the good. Bad decisions make us unhappy; but we are not disciplined enough to exercise our free will to choose only the good all the time.

2. Secondly, Adam and Eve are removed from the more intimate and immediate relationship they initially enjoy with God. In Genesis, this reality is depicted by having them physically driven out of paradise – another example of ingenious imagery communicating the essence of the truth. The word "paradise" means "within the walls" and the image in Genesis is of a garden surrounded by a wall inside of which Adam and Eve live in a special relationship with God and nature. Their bad choice means that they freely choose to rupture their special relationship with God and they are removed "outside the walls."

Thus, they, and we, lose this special relationship with God and begin to live with a more remote and less clear vision of Him. As with the consequence of internal strife, this consequence of a less-clear understanding of God and His loving relationship with each one of us exists in you. The upshot is that now you have to work hard to learn about God and His love for you. The fact that God lives and that He loves you is not intuitively known. You have to study and read and think and reason to find and know Him. Thus, it is good that you are reading this book; for it can be one of the tools you use to learn about Him and His great love for you, if you choose to do so.

3. Physical and mental weakness and vulnerabilities proliferate. Disease, effects of trauma, interpersonal discord, anxiety and individual and collective aggression and anxiety become part of our daily existence. We all become subjugated to nature's devastating events.

This is the ultimate reason why bad things can happen to people we see as good and not deserving of them. But let us examine this realty in a little more detail. First, none of us is really, essentially good – each one of us is well aware of how many times we have chosen to freely choose our baser impulses; hurting ourselves or others in the process. Second, as we will see in the next chapter, while Jesus gives us the opportunity to avoid the fourth consequence, described in the next paragraph, He does not remove all the consequences of Adam and Eve's sin. Therefore, suffering and pain is part of everyone's life on earth no matter how "good" a life someone is living.

4. The fourth, and most devastating, consequence is the loss of the opportunity to live forever in God's presence. Adam and Eve freely

choose to go their own way and to reject God's directive. As a result, they freely choose to forever live apart from Him. This consequence is also passed down to us as their descendants. Heaven, God's eternal realm, is out of bounds for all human beings. Each human being still has a human soul, and therefore has eternal life from the moment of conception when the soul is infused into the fertilized egg; but he or she no longer has the option of spending eternity with God in heaven until the death and resurrection of Jesus.

Theologians refer to Adam and Eve's bad choice as original sin and its four consequences as the fruits of that original sin. Others describe the inner strife consequence we all experience with psychological or psychiatric terminology. It is perhaps most useful, for purposes of this story, to do what is done above – to simply set forth the reality of original sin and describe its four consequences and to explain its cause as Adam and Eve's failure to do what God asked them to do.

In the end, they choose to utilize their highest faculties in some improper manner for their own purposes, and to not submit their free wills to God's directive, thereby denying Him His proper role in their lives. It is their pride that produces disobedience and the consequences of that failure live on in our own lives every day.

Now, of course, God could have left the matter here. But there is good news (i.e. Gospel) to report. God, our Creator, our Father, in His mercy, exercises His free will and creates a plan to ameliorate the ultimate horrible consequence of original sin that justice requires – that we humans be cut off from Him for all eternity.

He decides that He will have His Son, Jesus Christ, pay the price justice demands for this particular consequence, thereby providing each of us with another chance to live with Him in heaven, if we make good choices during our own sojourn here on earth.

The story of Jesus is discussed in relevant detail in the next chapter. For now, it is sufficient to note the following: when Jesus pays the price for our sins, we are not completely exonerated. If we are to take advantage of this, the greatest of all possible gifts, we must successfully contend with the other three consequences of original sin and strive daily to form an ever closer, more deeply loving, relationship with God. God does His part, freely and out of love. You and I now need to do our part if we are to live in an intensively intimate and loving relationship with Him in heaven.

After Adam and Eve are created, God directly creates every human being from that time to now by infusing a human soul into every human embryo. We know this must be the case, as we stated above, because we humans have self-awareness and we can abstract to the third level; and therefore we must have a spiritual soul that empowers us to create immaterial, spiritual concepts and such a faculty cannot evolve by mere materialistic causation.

With humans now present on the earth, steady incremental change in their geographical locations, individual relationships and organizational structures progress quite swiftly. What follows below is a timeline of the most significant events in our human history. The list demonstrates the accomplishments human beings have produced in just a fleeting moment in geographical timeframes.

Why is this relevant to your story? These events lead directly to our modern world and its social structures. Your values and judgments are formed, in significant part, by the physical and social environments in which you grow up and live. Thus, a review of the creative forces behind these environments is an important part of your story.

50,000 (approximate) B.C.[20] -- Standardized stone tools and jewelry (ostrich-shell beads) are crafted; art appears. As time passes, humans diversify into 6 broadly classified groups: blacks, Khoisan (Bushmen-Hottentots), African pygmies, whites, Asians and aboriginal Australians.

43,000 B.C. -- Watercraft are built with sufficient craftsmanship and strength that humans travel to, and occupy, Australia even though it is not visible from any previously occupied lands and requires navigating many deep-water channels up to 50 miles across.

20,000 B.C. -- Clothing (sewing) develops to the point that humans can live in Siberia.

---

20 Much of the timeline and its content are summarized from a most wonderful, interesting book: *Guns, Germs and Steel* by Jared Diamond. It has been summarized as a short empirical history of everybody for the last 43,000 years.

13,000 B.C. – Tribal organization (population of a few hundred) appears in the Fertile Crescent (defined on the next page).

12,000 B.C. – Pottery appears in Japan.

10,000 B.C. – Humans populate North and South America from Alaska to the tip of South America. All continents except Antarctica now occupied.

9,000 B.C. – Weaved cloth appears.

8,500 B.C. – Plant domestication of peas, wheat and barley.[21]

This initiates the rise of agriculture. The hunter-gatherer lifestyle is gradually eliminated with very few exceptions. (One group, the Australian aborigines, lasts up to very recent times.)

Domestication of plants and animals is a major accomplishment that is essential to creating sophisticated and complex civilizations up to, and including, our modern societies. Wondrously, it all came about in a very short period – about 500 years.

It occurs in the part of the world called the Fertile Crescent. Think of this area in the shape of a large boomerang with one end starting in Egypt, then passing north up through Israel, Jordon, Syria and into southern Turkey then rounding east and heading back down south through Iraq and western Iran ending at Kuwait and the Persian Gulf.

The Fertile Crescent is perfectly positioned geographically and climatically to create many of our civilizations' basic infrastructures (e.g., cities, writing and empires). But we also need to recognize that its inhabitants do a tremendous job with these resources and that we

---

21 As explained in *Guns, Germs and Steel*, wheat and Barley undergo a favorable change when a mutant gene prevents their stalks from shattering, a feature that had prevented earlier domestication. Plant domestication may be defined as growing a plant and thereby, consciously or not, causing it to change genetically to make it useful to consumers. Adoption of food production exemplifies an autocatalytic process, i.e., one that catalyzes itself in a positive feedback cycle, going faster and faster once it has started.

are much indebted to them for their accomplishments from which we greatly benefit today.

One of the Fertile Crescent's advantages over the rest of the world is the availability of edible plants. There are more than 200,000 species of wild flowering plants, but only a few thousand are edible and only a few hundred are domesticated. The Fertile Crescent's Mediterranean Zone has 32 of the world's 56 large-seeded grass species. Eight "founder crops" are domesticated in the Fertile Crescent – the cereals: emmer, wheat, einkorn wheat and barley; the pulses: lentil, pea, chickpea and bitter vetch; and the fiber crop flax. Only flax and barley have a large range in the wild outside the Fertile Crescent.

8000 B.C. Animal (sheep goats) domestication.

Humans and most animal species make an unhappy marriage for one or more reasons: the animal's diet, growth rate, mating habits, disposition, tendency to panic and/or various distinct features of its social structure may mitigate against close human contact. Domesticable animals, generally speaking, have 6 characteristics in common. (A domesticated animal is defined as one that is selectively bred in captivity and thereby modified from its wild ancestors for use by humans who control the animal's breeding and food supply.) These six characteristics are:

Sufficiently docile	Cheap to feed	Rapid growth
Submissive to humans	Immune to diseases	Breed well in captivity.

There are 148 species of wild large mammalian terrestrial land animals. None of them arise in four of the five worldwide Mediterranean Zones (Fertile Crescent, California, Chile, SW Australia and South Africa. The only such climate zone with some of these species is the Fertile Crescent, which has four of the five most important domesticable species of large mammals (i.e., 100 lbs or more) – goat, sheep, pig and cow. The only other large mammal to be domesticated and become important around the world is the horse. These large domesticated mammals have 3 characteristics:

- They live in herds;
- They have a well-developed dominance hierarchy among herd members; and

• Their herds occupy overlapping home ranges rather than exclusive territories.

As a somewhat humorous aside, cats and ferrets are the only territorial terrestrial mammals we have domesticated. We succeeded with these animals because our motives were different – we do not wish to gather cats or ferrets in large herds raised for food; but, rather, we keep them as solitary pets or hunters. Hence, the expression about a group of people who are difficult to control – "It is like trying to herd cats!" (As this is being written, the author has 2 cats and 2 ferrets and he cares deeply about them all; but he can testify that they do indeed resist herding.)

There is also a downside to animal domestication. The major killers of humanity in recent history are infectious diseases that come to us from our animal friends – smallpox (as far back as 1,600 B.C.), flu, tuberculosis, malaria, plague, measles and cholera.

7000 B.C. – Sugarcane domesticated in New Guinea.

5500 B.C. – Chiefdoms (populations from few thousand to a few tens of thousands) established in the Fertile Crescent. For the first time individuals learn how to encounter strangers without attempting to either establish some near or far blood relation with one another or kill each other. (Religion, by providing a common ideology, helps promote this ability.)[22]

4000 B.C. -- First fruit and nut trees domesticated – olives, figs, dates, pomegranates and grapes.

These plants prove more difficult to domesticate since they yield fruit only after 3 years and reach full production in 10 yrs.

4000 B.C. -- Horses domesticated in the steppes north of the Black Sea. Warfare is changed for centuries to come.

3700 B.C. – First state organization (population usually more than

---

22 The word "religion" probably comes from the Latin word, "religare" to bind fast (think of ligament). The concept is that through your religion you become bound fast to God and His will.

50,000) arises in Mesopotamia (Greek for "between rivers" – the central section of Iraq running NW to SE).

The overall historical trend has been for humans to organize themselves in increasingly complex social structures. One bad consequence – fanaticism in war is only possible in large groups. Smaller groups cannot afford to lose their young men so readily.

3400 B.C. – Wheel invented near the Black Sea.

3200 B.C. – Earliest writing invented by the Sumerians of Mesopotamia. (The creation of more complex writing using an alphabet will follow around 1,500 years later.) The main creative initiative in this development seems to be the need for a more efficient administrative tool to better keep track of enslaved human beings. The ability to read and write remained the exclusive province of the wealthy and relatively few highly learned individuals for tens of centuries.[23]

2500 B.C. – Camels are the last large terrestrial mammals to be domesticated.

A formal belief in an afterlife appears in Egypt. The Egyptians come to believe that after Osiris was killed, his wife, Isis, resurrected him with the Ritual of Life. This ritual is later given to the Egyptians so that all their dead may enjoy eternal life.

2300 B.C. – First empire (i.e., control over large area and many peoples). Sargon the First is the great ruler of the Akkadian Empire (located

---

23 Smaller societies tend to be hunter-gatherers and their societies failed to develop writing for two reasons:
1. There was no institutional use for it. They have no need for record keeping or political propaganda. In their culture they have an egalitarian political sphere confined to their own band's territory; and
2. They lack the social and agricultural mechanisms for generating food surpluses to feed individuals who could engage in learning and performing writing tasks.

Larger societies can have an efficient food-producing population. This leads them to need and create:
- Leadership and bureaucracies;
- Standing armies for wars of conquest; and
- Dispute resolution.

in parts of Mesopotamia, Syria and Iran). He pioneers techniques of imperial rule.

2000 B.C. – U.S. founder crops (those most easily domesticated) established: squash, sunflowers, sumpweed (related to the daisy) and goosefoot (related to spinach).

1850 B.C. – Abraham ("father of many") called by God to found a people that He will use in His plan to send His Son Jesus to pay the price for Adam and Eve's sin. As a foreshadowing of God the Father's sacrifice of His own Son Jesus, Abraham is willing to sacrifice his son Isaac (Genesis 22:1-24). This is the beginning of the Jewish people and the Jewish religion.

1700 B.C. – Printing invented in Crete (the Phaistos disk) and then forgotten until 750 A.D. in China and 1400 A.D. in Europe (gaps of more than 2,400 and 3,100 years respectively).

1550 B.C. – Consonant alphabetic writing created in the western part of Fertile Crescent.

1500 B.C. -- Hinduism starts in India. Unlike other major religions, it does not have a single founder, a specific theological system, a single system of morality, or a central religious organization. It consists of thousands of different religious groups, mostly still located in India. It is now the world's third-largest religion, after Christianity and Islam, and claims about 837 million followers - 13% of the world's population.

Strictly speaking, most forms of Hinduism are henotheistic. This means believers recognize a single deity who has a special relationship with them or their family; but they also recognize other gods and goddesses as facets, forms, manifestations or aspects of that supreme God. Hindus believe in the repetitious Transmigration of the Soul -- the transfer of a person's soul after death into another body. They believe we experience a continuing cycle of birth, life, death and rebirth through their many lifetimes.

1350 B.C. – Moses delivers God's 10 commandments and establishes the Jewish law.

1000 B.C. – King David writes Psalm 22 that graphically, with great accuracy, details the agony and torture Jesus will undergo as He suffers through His passion and crucifixion a thousand years later.

800 B.C. – Vowels invented by the Greeks. (Opened up reading and writing to many more people.)

624 B.C. -- Buddha Shakyamuni is born. "Shakya" is the name of the royal family into which he was born and "Muni" means "Able One. He sets out to find enlightenment through meditation. He attains varja-like concentration that he believes is the highest mental state of a limited being. In this state, he removes the final veils of ignorance from his mind and in the next moment becomes a Buddha -- a fully enlightened being. Buddhists believe we need to meditate each day and train our minds to think only constructive thoughts.

551 B.C. – Confucius is born. His philosophy emphasizes personal and governmental morality, correctness of social relationships, justice and sincerity. This value system comes to predominate over other faiths and doctrines in China.

500 B.C. – Cast-iron production in China.

490 B.C. – Persian wars. Darius defeated at Marathon and Xerxes weakened at Thermopylae (480 B.C.) by the famous 300 Spartans led by Leonidas. This halts Persia's invasion into the west.

400 B.C. -- The Greeks develop two principles to live by:
1. An individual is responsible for his own life and his own actions; and
2. The afterlife is available to anyone from king to farmer.

325 B.C. -- Aristotle creates his theory of duality of act and potency. Act is form and potency is matter. In living beings, we use a specialized

terminology for these realities -- act is the soul and potency is the body.[24]

221 B.C. – China achieves political unity under the Qin Dynasty.

2 B.C. Jesus Christ is born

---

24 There is a story about how his theory of duality came about. One day Aristotle goes to a friend's funeral. As he looks upon his friend's dead body, he begins to wonder how he is different today from when he was alive just yesterday. His body looks intact; but whatever gave that body life, whatever made it alive, was gone.

From these musings Aristotle eventually reaches the brilliant conclusion that his friend had to consist of two components – a material, passive component, the stuff from which his body was made and a non-material, active component which coherently organized ("informed") his friend's body, rendering his friend a living human being. This insight became known as "the principle of duality" and it asserts that all living entities consist of two components – a material structure and a soul – including plants, animals and human beings.

This principle of duality becomes a basic tenet of western civilization's philosophy and is an integral part of our culture today. This theory applies to the whole material universe. The tree in your yard has a body and soul but its soul does not perform any spiritual functions and therefore does not require the direct intervention of God when the tree starts its life, nor does it survive when the tree dies.

CHAPTER IV

THE ARRIVAL, LIFE AND DEATH OF JESUS CHRIST

Therefore, the Lord himself will give you this sign: the virgin shall be with child, and bear a son, and shall name him Immanuel. Isaiah 7:14 (c. 750 B.C.)

Jesus is born in Bethlehem into a poor family with little material resources. His mother is a Jewish teenager named Mary who, through the miraculous power of God, conceives her son while remaining a virgin. His stepfather is Joseph, a carpenter and a descendant of the House of David.

In spite of Jesus' lowly social and economic status, our modern history is divided into two eras by His birth. More than 1,500 years ago, a Roman monk, Dionysius Exiguus, created the term *Anno Domini*, abbreviated A.D., "in the year of our Lord" for the time that has passed since His birth. The abbreviation B.C. "before Christ" is for time before His birth.

There is now a growing effort to try to divorce Christ from this historical division of our date calculations by using the abbreviations C.E., "common era," in place of A.D. and B.C.E., "before the common era" for B.C. However, note that, in fact, these new designations still relate to our historical dating system, and therefore both are inexorably tied to Christ's birth, even though He is not directly referenced in these new designations.[25]

---

25 Dionysius probably erred in establishing the year of Jesus' birth – it is most likely to have been 2 B.C. on our calendar. But the point is His

Christians believe that Jesus is the Son of God sent by His Father to save us from our sins. This chapter presents a summary of His mission and discusses why Christians believe as they do and what this belief means for anyone who accepts it.[26]

There have been men, from time to time, who claim they have been sent by God to reveal His truths to their fellow human beings. It would seem to be a reasonable expectation that if God were to send such a messenger he would be pre-announced before he ever arrived on the earth. After all, this advance notice is practiced by our commercial enterprises all the time. Apple, Microsoft, General Motors, Toyota, etc., all put on annual events, at great expense, to announce their new products for the coming year.

Surely, God would let us know in advance if He were going to send a special envoy to us. Even more so, would this not be expected if the messenger were to be God's own Son? Absent such a record, how would we distinguish the legitimate messenger from the fraudulent one?

It turns out, looking back through history, that God did precisely this. There is one messenger claiming to be sent by God, and only one, who was announced in advance. Jesus' arrival was foretold hundreds of years ahead of time, throughout the whole of the known world, and on many separate occasions. These announcements are astounding in their detail, numerousness, length of time over which they are delivered and geographical spread.

They include such details as where He would be born, facts about His mother, Father and stepfather, where He would live, the doctrine He would teach, the enemies He would make, the life-program He would adopt for our future, and the manner of His death. What follows is a summary presentation of some of the prophecies that foretold many details about the life of Jesus. They are each set forth, in a numbered section, in this format:

(Footnote 25 continued.) birth divides our entire dating system into two distinct periods, no matter the exact date of His birth.

26 Outside of the Gospels themselves, the best book on the life of Jesus, written to date, is *The Life of Christ* by Bishop Fulton J. Sheen (New York: Reprinted by Image, 2008). Many of the concepts and insights I offer in this chapter have their genesis in this work. I highly recommend it to any reader who has a genuine interest in Jesus and wonders what His life means for him or her.

- The prophetic announcement is paragraph "A";
- The date of the prediction or the mid-age of the person who makes it is paragraph "B"; and
- A description of how it is fulfilled in Jesus is paragraph "C."

It is an impressive collection, to say the least, and it certainly contains the reasonable level of advanced notice that we would expect from God if He were to send us a special messenger.

The Jewish advance notices[27]

    1. The announcement. God announces that a descendant of the Patriarch Abraham will be the instrument through which all of us will be blessed.
    1.A. Genesis 12:3[28] --

The LORD said to Abram: 'Go forth from the land of your kinsfolk and from your father's house to a land that I will show you.
'I will make of you a great nation, and I will bless you; I will make your name great, so that you will be a blessing.
'I will bless those who bless you and curse those who curse you.
All the communities of the earth shall find blessing in you."
(Emphasis added.)

Genesis 12:1-3

    1.B. Abraham lived around 1850 B.C. Tradition tells us that Moses wrote Genesis around 1350 B.C.
    1.C. First, Christians believe that Jesus died to save all human beings from their sins so all are blessed in Jesus. Second,

---

27 In all, there are 113 Messianic prophecies about, or events foretelling, the Messiah in the Jewish scriptures. Jesus, and He alone, fulfills every one of them with exactitude. To read about the astronomical signs in the sky around the time of Jesus' birth – December 2 B.C. (and His death on the cross – April 3, 33 A.D.) go to http://www.bethlehemstar.net and read the fascinating, high-quality scientific research and analysis done by Frederick A. Larson. He demonstrates, with facts that are startling in their detail, that even the stars and moon dramatically acknowledged Jesus' birth (and death).

28 All biblical quotations are from the *New American Bible,* Conference of Catholic Bishops, Washington, DC, 2002, unless otherwise noted.

Christians, who are the spiritual descendants of Abraham, now reside in every country, and in virtually every community, in the world and produce blessings for their neighbors that are beyond measurement.

2. **Advance detailed notice about how His messenger was to live and die.**
   2.A. Psalm 22: 2-19 –

My God, my God, why have you abandoned me? Why so far from my call for help, from my cries of anguish?
My God, I call by day, but you do not answer; by night, but I have no relief.
Yet you are enthroned as the Holy One; you are the glory of Israel.
In you our ancestors trusted; they trusted and you rescued them.
To you they cried out and they escaped; in you they trusted and were not disappointed.
But I am a worm, hardly human, scorned by everyone, despised by the people.
All who see me mock me; they curl their lips and jeer; they shake their heads at me:
'You relied on the LORD--let him deliver you; if he loves you, let him rescue you.'
Yet you drew me forth from the womb, made me safe at my mother's breast.
Upon you I was thrust from the womb; since birth you are my God.
Do not stay far from me, for trouble is near, and there is no one to help.
Many bulls surround me; fierce bulls of Bashan encircle me.
They open their mouths against me, lions that rend and roar.
Like water my life drains away; all my bones grow soft. My heart has become like wax, it melts away within me.
As dry as a potsherd is my throat; my tongue sticks to my palate; you lay me in the dust of death.

Many dogs surround me; a pack of evildoers closes in on me.
So wasted are my hands and feet
that I can count all my bones. They stare at me and gloat;
they divide my garments among them; for my clothing they
cast lots.
(Emphasis added.)

2.B. King David, the author of this precisely detailed prophetic Psalm, lived 1,000 years before Jesus was born.
2.C. A very accurate description of the way Jesus was treated and the effects of crucifixion on the body. As blood drains from the body, an extreme thirst sets in, one that is beyond description. This text capture that terrible ordeal in most vivid imagery.

3. **The Bible of the Alexandrian Jews, the Septuagint, predicts a virgin birth.**
   3. A. Isaiah 7:14 --

   Therefore the Lord himself will give you this sign: the virgin shall be with child, and bear a son, and shall name him Immanuel.

   3.B. Isaiah lived 750 B.C.
   3.C. Christians and Muslims believe that Jesus was born to a virgin named Mary.

4. **Isaiah describes His "normalness," His "ordinariness" and His extreme suffering for us.**
   4.A. Isaiah, Chapter 53: 1-12

   Who would believe what we have heard? To whom has the arm of the LORD been revealed?
   He grew up like a sapling before him, like a shoot from the parched earth; There was in him no stately bearing to make us look at him, nor appearance that would attract us to him.
   He was spurned and avoided by men, a man of suffering, accustomed to infirmity, One of those from whom men hide

their faces, spurned, and we held him in no esteem.

Yet it was our infirmities that he bore, our sufferings that he endured, While we thought of him as stricken, as one smitten by God and afflicted.

*But **he was pierced for our offenses**, crushed for our sins, Upon him was the chastisement that makes us whole, **by his stripes we were healed.***

We had all gone astray like sheep, each following his own way; But the LORD laid upon him the guilt of us all.

Though he was harshly treated, he submitted and opened not his mouth; Like a lamb led to the slaughter or a sheep before the shearers, he was silent and opened not his mouth.

***Oppressed and condemned, he was taken away**, and who would have thought any more of his destiny? When he was cut off from the land of the living, and smitten for the sin of his people,*

***A grave was assigned him** among the wicked and a burial place with evildoers, Though he had done no wrong nor spoken any falsehood.*

(But the LORD was pleased to crush him in infirmity.) If he gives his life as an offering for sin, he shall see his descendants in a long life, and the will of the LORD shall be accomplished through him.

*Because of his affliction he shall see the light in fullness of days; **Through his suffering, my servant shall justify many, and their guilt he shall bear.***

Therefore I will give him his portion among the great, and he shall divide the spoils with the mighty, Because he surrendered himself to death and was counted among the wicked; And he shall take away the sins of many, and win pardon for their offenses. (Emphasis added.)

4.B. These verses precede His birth by 750 years.

4.C. This description fits the life, passion and death of Jesus in specific detail. Note the bold sections that predict that He will be pierced (crucified) be given stripes (the scourging), acquiesce completely to His passion and death to save us from our sins and be placed in an assigned grave (given to Him by Joseph of Arimathea).

5. **The Book of Jonah is written around 500 B.C. The story foretells Jesus spending three days in the "belly" (grave) of the earth.**

6. **Also, about 500 B.C., Zechariah's prophecies speak with detailed accuracy about Jesus' entrance into Jerusalem on the day that has become known as Palm Sunday, the price paid to Judas for his betrayal and Judas flinging the coins back into the Temple after he realizes he has betrayed an innocent Jesus.**

> *Rejoice heartily, O daughter Zion, shout for joy, O daughter Jerusalem! See, your king shall come to you; a just savior is he, Meek, and riding on an ass, on a colt, the foal of an ass.*
>
> Zechariah 9:9

> *I said to them, 'If it seems good to you, give me my wages; but if not, let it go.' And they counted out my wages, thirty pieces of silver.*
> *But the LORD said to me, 'Throw it in the treasury, the handsome price at which they valued me.' So I took the thirty pieces of silver and threw them into the treasury in the house of the LORD.*
>
> Zechariah 11:12-13

Pagan Advance Notices

7. **Chinese authors made similar predictions.**

   7.A. *In the 24th year of the reign of Tcheao-Wang of the dynasty of the Tcheou, . . . on the 8th day of the 4th moon, a light coming from the South-west which illumined the palace of the king. The monarch, beholding this splendor, interrogated concerning it the sages. They presented to him books wherein it was written that this prodigy would announce that a great Saint had appeared in the West, and that in a thousand years after his birth his religion would be spread into those parts.*[29]

---

29 Huc, Evariste Regis. *Travels in Tartay, Thibet, and China*, Abington: RoutledgeCurzon, 1928, 2005. p. 87. Print.

7.B. The year was 1029 B.C.

7.C. China was on the other side of the world. That is why the prophecy relates to a great Wise Man, a founder of a great religion, who would be born in the West.

8. **Confucius (See above on page 28.)**
   8.A. He spoke of "*the Saint*" who was to come, probably reflecting the prophecy related above.
   8.B. He lived 520 B.C.
   8.C. The Saint is a person who would live in God's presence and do God's will.

Advance Notices from Roman Authors

9. **Cicero, the most famous Roman orator also gave a firm prediction.**
   9.A. *After recounting the sayings of the ancient oracles and the Sibyls* [ancient prophetesses] *about a 'King whom we must recognize to be saved,' asked in expectation, 'To what man and to what period of time do these predictions point?*[30]
   9.B. He lived in 70 B.C.
   9.C. This is the first Roman connection between accepting a King and receiving salvation. The name Jesus ("He Who Saves") Christ ("The Anointed One") encompasses both concepts – salvation and lordship – and perfectly describes His earthly mission as prophesized here.

10. **Virgil, the most famous Roman poet, gives reinforcement to Cicero's words.**
    10.A. The Fourth Eclogue of Virgil recounted the same ancient tradition as Cicero and spoke *of a chaste woman, smiling on her infant boy, with whom the Iron Age would pass away.*[31]

    10.B. The Eclogues were written in 37 B.C.

---

30 Sheen, op. cit., p. 259.
31 Ibid., p. 260.

10.C. Again, we see the notion that a baby would be born to a woman of purity and He would change the world. Christianity produced a literate culture and this lead to the end of the Iron Age in various societies as it took hold, just as Virgil predicted.

11. Tacitus, a Roman historian, recounts a long-held belief among the Romans.

11.A. *People were generally persuaded in the faith of the ancient prophecies, that the East was to prevail, and that from Judea was to come the Master and Ruler of the world.*[32]

11.B. Tacitus lived 85 A.D.

11.C. Note how precise this ancient persuasion was – it pinpointed a small, insignificant Roman province as the location of the ONE who was to come.

12. Suetonius, another Roman historian, in his account of the life of Vespasian, recounts a Roman tradition thus.

12.A. *It was an old and constant belief throughout the East, that by indubitably certain prophecies, the Jews were to attain the highest power.*[32] He also quoted *a contemporary author to the effect that the Romans were so fearful about a king who would rule the world that they ordered all children born that year to be killed - an order that was not fulfilled, except by Herod* around 2 A.D.[33]

12.B. Suetonius lived 100 A.D.

12.C. The Jewish people were targeted by ancient rulers out of a fear, based on a widely spread belief, that they were to eventually produce the ruler over all the earth.

The Greeks also expected Him

13. Aeschylus, the premier Greek dramatist, was also aware of God's special messenger.

13.A. He writes in his *Prometheus Bound*: *Look for no term to such an agony till there stand forth among the Gods one who shall take upon him thy sufferings and consent to enter hell far from the*

---

32 Ibid., p. 260.
33 Ibid., p. 247.

light of Sun, yea, the deep pit . . . for thee.

13.B. He lived 490 B.C.

13.C. Here is the first pagan expression, and a very clear one too, of the counter-intuitive prediction that the expected special messenger from on high would be a suffering servant. This prediction would find its fulfillment in the suffering of Jesus.

14. Plato and Socrates spoke of "Logos" -- the Universal Wise Man "yet to come."[34]

When one reads all these "advance notices," it is obvious that God alerted peoples of the earth to the fact that He was going to send us a unique messenger and savior. He spent 2,000 years preparing the Jews as a special people who would provide the spiritual, ancestral and cultural parameters from which His Son could be made incarnate; becoming a Person Who is both true God and true man.

This messenger is willing to suffer for us, to carry our burden of sin; and, in the end, to die for us so that we might be saved. This special envoy is very different from us – we are born to live; He is the Son of God and already has eternal life. He is born for one purpose -- to die.

So Jesus arrives. His birth occurs on the outskirts of the small town of Bethlehem ("House of Bread") in a cave used to shelter animals because there is no room for Him in the inn. Many visitors have traveled to town to register pursuant to Caesar Augustus's decree for a worldwide census count and they have taken all of the town's lodgings.

He is born to a young Jewish teen-ager, the Virgin Mary. She freely assents to God's invitation, delivered to her by the Angel Gabriel, to collaborate with the Third Person of the Blessed Trinity, the Holy Spirit, to incarnate the Second Person of the Blessed Trinity, The Son, as a human baby boy. While His duality as true God and true man is unique in history, it is also somewhat possible for us to conceptualize it since we ourselves are a duality of soul and body, spirit and matter. Jesus is likewise a duality – a joining of the Godhead Son with human flesh and blood provided by Mary.

Shortly after His birth, Mary's husband, Joseph, takes the family

---

34 Ibid., p. 262.

and flees to Egypt for a few years, to escape King Herod the Great's attempt to pre-emptively kill the baby before He can grow up to become the prophesized king and threaten his own rule. As so often happens in the life of Jesus, people who react to, or interact with, Him get it very wrong. So is the case with Herod. He misses the whole point. Jesus was not born to establish a kingdom on earth; but to bring the Kingdom of Heaven into being by dying to save us from our sins. He is Herod's savior, not his destroyer.

When the family returns from Egypt, they settle in Nazareth, a small town about 75 miles north of Jerusalem, and Jesus lives is obscurity for the next 30 years, practicing the carpenter trade of His stepfather, Joseph. During all these years, we know of no particularly strong impression He makes on His neighbors.

In fact, with only one exception, we are aware of no significant event in His life during this entire time. The one exception occurs at the age of 12. He spends three days in the Temple in Jerusalem discussing religion and Jewish law with the intellectual and religious leaders of the day and He amazes them with His knowledge and responses.

During this discussion with the Jewish intelligentsia, Jesus must have raised the issue of a Suffering Messiah so graphically presented in Psalm 22 (page 51, supra) and Isaiah 53 (page 52, supra). Read this material again, slowly. Then consider, if you are 12 years old and know that these terrifying descriptions apply specifically to you, how you would be impacted by recurring thoughts of the torture-filled, agonizing hours that will precede your death.

Jesus, being truly man, must think about these prophecies with great dread and anxiety. Such thoughts must also be a daily reality for Him. For He is a carpenter and He works with wood and nails. He naturally contemplates, frequently, the role that wood and nails are to play in His death even as He works with these materials to help support His family. Yet, despite these horrific images, played repeatedly in His mind, He perseveres in His mission to suffer and die for our sins out of His great love for each one of us.

One wonders why God's plan for our redemption includes a 30-year period when His Son lives and toils obscurely in a small town in northern Israel making simple furniture, ox yolks, etc. The reasons for all these years of obscurity are, of course, ultimately, a mystery. But perhaps God had some of these lessons in mind with this period of obscurity:

- Humility;
- An honest, quiet, unassuming life; and
- All honest work has value. What you do for a living, specifically, is not important; what is important is that you do your job to the best of your abilities.

In any event, the fact is that we know very little about the first 90 percent of His life. We do know a great deal about the last three years thanks to the extensive details about Him in the 27 books of the Bible's New Testament and the church organization, the "ecclesia", that starts up 50 days after His departure from earth and preserves many oral traditions about Him that are passed on down through the centuries.

When He is around 30 years old, the time arrives for His mission to begin its public phase. His mission is not what most of His contemporaries expect of the Messiah Who is to come. He is not here to become king of Israel and defeat all of its enemies and begin its supremacy over all the peoples on the earth. His mission is radically different than this. It is to suffer and die for the sins of all people in order to give each one of us the opportunity to live with Him in heaven for all eternity if we each respond to God's graces and make good choices in our own lives.

The first public "announcement" is an unintended one. Jesus is at a wedding feast in Cana, a small town about five miles north of Nazareth in Galilee, when the wine runs out. A good host, in those days, provided all the wine any guest would desire; serving the best wine first and then the inferior later (when the guests would not notice!). At this feast, the host did not plan well, or did not have the funds to buy all that was needed, and the wine runs out. This is an embarrassing situation for the young couple. Mary, Jesus' mother, learns of this and asks Him to help. He tells her that it is not yet time for Him to begin working miracles and become a public figure. But she knows He will not let the young couple be embarrassed. The result is that He turns 150 gallons of water into high-quality wine.

Jesus now begins His mission in earnest. He gathers to Himself 12 men, His apostles ("those who are sent"), who are to be with Him for the rest of His life on earth. To them will be entrusted, eventually, the

Good News, with the responsibility that they carry His message to all the peoples in the world.

Over the next three years He travels all over Galilee and Judea, working thousands of miracles,[35] talking about the Kingdom of Heaven and teaching about the relationship that exists in the Godhead among The Father, Himself as the Son, and the Holy Spirit. He slowly discloses to His apostles the reason He has come – to offer Himself up to suffering and death as the one sacrifice that can pay the price for our sins and allow us potential access to heaven.

He teaches that God is a God of love Who desires that every human being live his or her life also as one of love. He teaches that all the Jewish commandments and laws can be boiled down to just two supreme commandments that apply to every human being:

• Love God with all your heart, soul, strength and mind; and
• Love you neighbor as yourself.

To each one of us, who lives by these two commandments, He promises eternal life with His Father, the Holy Spirit and Him in heaven, living in perfect harmony and happiness without end. Such an obedient person's soul, upon departing from this world, in the process we call death, goes to judgment before Jesus where he or she will be found worthy, or not worthy, to live for all eternity in the presence of Love Personified, communing with God face-to-face.

He also understands that very few of us can live such a life from start to finish. He tells us to ask for His help in time of temptation and that we need to seek forgiveness when we fail to live by these two commandments.

THE key to life, bar none, is to live in obedience to these commandments AND to get yourself up off the floor when you fail to do so and go to God in sorrow and sincere repentance and start over, re-

---

35 We do not know exactly how many miracles Jesus performed. St. Matthew and St. Luke each record 20 specific miracles, St. Mark, 18 and St. John 7. But these writers make several general statements such as "He worked many other signs" or "He cured the blind," etc. St. John ends His Gospel thus:

There are also many other things that Jesus did, but if these were to be described individually, I do not think the whole world would contain the books that would be written.

ST. JOHN 21:25

dedicated to living more faithfully tomorrow. You cannot overestimate the forgiving love of God so long as you are doing your part -- trying to steadfastly live by these two commandments in good faith and honestly seeking forgiveness when you fail to make good choices and do not act out of love.

Based on the crucial importance of these two commandments, the key question becomes: exactly who is my neighbor whom I am supposed to love? Jesus tells us precisely what He means by the word "neighbor" in this wonderful story.

> *Jesus replied, 'A man fell victim to robbers as he went down from Jerusalem to Jericho.*
> *They stripped and beat him and went off leaving him half-dead.*
> *'A priest happened to be going down that road, but when he saw him, he passed by on the opposite side.*
> *'Likewise a Levite came to the place, and when he saw him, he passed by on the opposite side.*
> *'But a Samaritan traveler who came upon him was moved with compassion at the sight.*
> *'He approached the victim, poured oil and wine over his wounds and bandaged them.*
> *Then he lifted him up on his own animal, took him to an inn and cared for him.*
> *'The next day he took out two silver coins and gave them to the innkeeper with the instruction, 'Take care of him. If you spend more than what I have given you, I shall repay you on my way back.'*
> *'Which of these three, in your opinion, was neighbor to the robbers' victim?'*
> *He answered, 'The one who treated him with mercy.' Jesus said to him, 'Go and do likewise.'*
>
> Luke 10:30-37

To fully understand this parable, you need to have extensive background information. There are particular experiences and customs peculiar to this time and place that provide a rich context for each verse and they must be known for this story to convey its full meaning.

A "man" went down from Jerusalem to Jericho. His exact identity is not essential and is therefore not detailed, but he most likely is a Jew. What Jesus does emphasize is the victim's condition -- not his nationality, religion, education, wealth, social standing or physical appearance. He is simply a man -- a fellow human being who eventually comes to need help.

Jericho is 3,300 feet lower in elevation and about 17 miles northeast of Jerusalem if one travels along "The Way of Blood," so called because of the numerous criminals who prey on its travelers. Jesus does not belittle the man for making so dangerous a journey alone – a foolish decision by any standard. He does not intend to create an excuse for those who refuse to help by criticizing the victim's poor decision to take this route and to do so alone.

This man "fell victim to robbers." The Greek word used means that these criminals encircle him. The robbers of Jesus' day were often violent and they had no hesitation about inflicting serious bodily harm on their victims. And, there were lots of these desperate men around at this time. Josephus writes that out of some 40,000 men dismissed by Herod the Great from building the temple a large number of them turned to this violent occupation.

"A priest happened to be going down that road . . . . Likewise a Levite came to the place . . . ." There were about 12,000 priests and Levites residing in Jericho at this time and they would travel back and forth on this road to their jobs in the temple. Due to their official status in the temple, the local hooligans on this road did not attack them.

Note the reaction of these two religious leaders. First, the priest "... when he saw him,   . . . passed by on the other side." Then, the Levite "... came and looked on him, and passed by on the other side." They both saw the man, knew there was a serious need for help, and yet they both went over to the other side of the road and kept on going. They felt guilty enough to get as far away physically as they could, but not empathetic enough to render even the most rudimentary assistance. One is left to wonder how many other times these two religious officials had seen travelers in distress and ignored their needs and walked "on the other side" without helping.

"But a Samaritan traveler who came upon him was moved with compassion at the sight." With this statement, Jesus really draws a line in the sand with his Jewish audience since the Samaritans and the Jews had been at each other's throats for over 700 years.

In 722 B.C., the Assyrians conquered Samaria, home to the 10 northern tribes of Israel. Many of the captive Jews were taken back to Assyria and in their place a foreign upper class of people were imported from Babylon and other places. This resulted in the formation of a hybrid race, which came to be the Samaritans.

From that time onward, there was never a feeling of kinship between the Samaritans and the remnant of Jews living in the southern kingdom of Judah. The Jews who resettled Jerusalem after the captivity considered the Samaritans as mongrels or half-breeds and not Jewish. In fact, when one Jew sought to degrade another in one of the strongest ways possible, he would call him a "Samaritan."

So, in this story, Jesus really hits each of His Jewish listeners right between the eyes by making the generous rescuer a Samaritan. Moreover, Jesus says: "...when he saw him, he had compassion on him." The Greek word used for "compassion" is a strong one and means, "his heart went out to him." The priest and Levite lacked the compassion that the Samaritan possessed.

True compassion always manifests itself appropriately. In this case, the Samaritan uses what he has with him -- oil and wine to treat and cover wounds, his beast and funds to pay for care and lodging -- to assist the injured traveler.

When Jesus asks his questioner, "a scholar of the law," which of these three passers-by was a "neighbor" to the victim, the lawyer has no answer, save one, the Samaritan. But he simply cannot bring himself to give that precise response, so he states, obliquely: "the one who treated him with mercy." Jesus then says to him, and to each one of us, that we are to "Go, and do thou likewise."

As you think about this short story, there are three philosophies of life that manifest themselves.

1. The philosophy of the robbers is completely self-centered. Only their lives matter – the rest of mankind is there to fulfill their needs no matter what it costs other people.
2. The philosophy of life practiced by the priest and Levite is that while we should not cause direct harm to anyone, each one of us is to look out for only himself or herself. If someone else has an unfortunate accident; or comes down

with a serious, debilitating illness, or the like, that is too bad; but that is not my problem. They committed the all too common sin of omission in failing to assist another wounded human being.

3. In stark contrast to both of these philosophies, the "Good" Samaritan makes time to help and uses what he has on hand to assist the traumatized victim, without counting the cost.

In sum, all three philosophies of life are presented here:

- What is yours is mine;
- What is mine is mine; and
- What is mine is yours whenever you need it.

So when Jesus tells his audience that there are only two commandments – love God with your whole being and love your neighbor as yourself, he leaves no room for guessing what He means. We are to love God and our neighbor – anyone we meet on the way through life who needs assistance – unconditionally and with generosity. We do not judge and we do not second-guess; we are to just love, serve and share. Of thus, are the citizens of the Kingdom of Heaven.

Sadly, many of us refuse to obey these two commandments and decide instead to live for the self instead of for God and for our neighbors. We do not want to obey these moral imperatives because it means giving up various selfish and sinful pleasures to which we have grown attached. So we close off our minds, shut down our consciences, and go down the broad avenue that leads to eternal misery.

It is simply amazing how resistive we can be; to what extent we can harden our human hearts to the truth. We can freely choose that our heart grow so hard that nothing, not even God Himself, can break through the barrier. Two examples of how hard a heart can be come to mind.

The first example demonstrates that many miracles, even when performed by Jesus Himself, are not enough to break through barriers some individuals put up.

Then he began to reproach the towns where most of his mighty deeds had been done, since they had not repented.
'Woe to you, Chorazin! [a town in Galilee, 2.5 miles north of Capernaum]
Woe to you, Bethsaida! [3 miles east of Capernaum, across the Jordon

River] *For if the mighty deeds done in your midst had been done in Tyre* [gentile city 30 miles NW of Capernaum on the Mediterranean Sea coast] *and Sidon* [gentile coastal city 20 miles NE of Tyre], *they would long ago have repented in sackcloth and ashes.*

'But I tell you, it will be more tolerable for Tyre and Sidon on the day of judgment than for you.

'And as for you, Capernaum [Peter's hometown in Galilee]*: Will you be exalted to heaven?*

You will go down to the netherworld. For if the mighty deeds done in your midst had been done in Sodom, it would have remained until this day.

'But I tell you, it will be more tolerable for the land of Sodom on the day of judgment than for you.'

Matthew 11: 20-24

Second, there is the case of Judas Iscariot, one of the original twelve apostles, who was with Jesus night and day for three years. He saw and heard His preaching, witnessed His generous response to all in need, including His innumerable miracles; yet, he choose to betray Him with a kiss the night He was arrested and condemned to death.

This point cannot be overemphasized. The hardness to which the human heart can calcify defies description. Many human beings who --

- Interacted with Jesus face-to-face;
- Witnessed His love and His kindness to anyone who asked for help;
- Listened to Him speak gently and convincingly of the Kingdom of Heaven and about the loving Father we all share; and
- Repeatedly witnessed his miraculous cures and commands obeyed by nature itself. --
- Rejected His personal call to them to give up their lives of sin and disharmony and follow Him in love.

Jesus never rejected anyone who came with an open mind to listen to Him; He never harmed anyone and did only good deeds His entire life. Yet, He became so hated that, the day He died, He was physically brutalized beyond recognition and put to death in the most painful, shameful manner available at the time – crucifixion. This is the power sin can acquire over hardened human hearts.

In spite of all this rejection, which Jesus experienced during His lifetime on earth and continues to experience today from heaven – for He is still calling each one of us to leave our life of sin and to come, follow Him – when anyone of us chooses to ignore His call and fails to live and love as we should, He continues to love us and to remain present with His hand held out to lift us up the moment we repent and turn to Him for help in walking the narrow path of righteousness.[36] So great is His love for each one of us, that He never abandons us; it as always we who choose to leave Him.[37]

No one can minimize how difficult it is to love, as commanded by Jesus. It takes a real commitment, renewed every day, to try to live up to this standard. But we must try each day for this is what we all are called to do.

The good news is that Jesus is standing by; ready to offer assistance in our effort to so live our lives, if we ask Him. He paid the price for our sins with His death on the cross and He showers us with grace every moment of every day if we but open our hearts to receive it. If we pray to Him for help, we can live in obedience to these two commandments. If we do not pray, and try to do so on our own, it is much more difficult, if not impossible.

Jesus' assistance is possible because He was faithful to the Father's plan to redeem us. As I said above, Jesus came into this world to die. No human being could pay the price to redeem humanity from Adam

---

36 This interior battle between our better and our sinful self is as old as humanity itself. More than sixteen hundred years ago St. Augustine, who lead a dissolute life for many years before turning his life around and becoming one of the Roman Catholic Church's greatest theologians, wrote a book entitled *Confessions*. In it he describes this war: *Ingrained evil had more hold over me than unaccustomed good. The nearer approached the moment of time when I would become different* [i.e. start living a life of selfless love], *the greater the horror of it struck me. But it did not thrust me back nor turn me away, but left me in a state of suspense.* Kindle Edition, Location 2547-53.

37 *You care for every one of us as if You care for the individual only. You care for all as if there were but one!* Saint Augustine, *Confessions*, Florida: Bridge-Logos, 2003. p. 70. Print. This means that even if you were the only human being ever created, and you sinned, Jesus would die for you on a cross, so great is His love for you.

and Eve's fall from grace. But Jesus, since He is both true God and true man, can do so if He is willing to pay the price on our behalf. Only His sacrifice could satisfy God's attribute of Justice, thereby affording each one of us the opportunity to live forever with Him in the Kingdom of Heaven if we obey His two commandments of love.

The price Jesus paid was high – a testament to how serious sin is. The best visual depiction of Jesus' arrest, torture and death is in the movie The Passion of the Christ.[38] It is historically accurate and it helps us understand how tremendously courageous, dedicated and loving was Jesus to voluntarily go through this awful ordeal in order to save those of us who choose to live our lives in His love.

His Passion consists of five different tortuous experiences. It begins with a visit to the Garden of Gethsemane, a peaceful place located at the foot of the Mount of Olives immediately east of Jerusalem. Jesus often went here to pray.

On the night He is arrested, Jesus leaves Jerusalem, immediately after celebrating the Passover supper with His disciples (and, in Roman Catholic tradition, instituting the Mass and Holy Eucharist celebration) and walks to this Garden. He knows this is the night He is going to be tortured and that He will be crucified in the morning just as Sacred Scripture has foretold.

For several hours, He goes through a horrible internal ordeal, referred to as the agony in the garden. Several times He prays to His Father, saying that He, the Father, can do all things and that, if it be possible, He asks His Father to "let this cup [His passion] pass from Me". But then He always ends this prayer, in a display of remarkable courage: "Not my will but Yours be done."

What causes Jesus such distress will never be known exactly. He knew for many years that He was born to die as a suffering messiah. So, it may well be that this severe internal angst is caused by much more than concern over the physical suffering He is about to undergo, as horrifically brutal as that will be. (He had to have seen victims of scourging and crucifixion and knew what to expect.)

I believe His agony is much deeper than that. Such severe distress (St.

---

38 I highly recommend this movie for adults. (It is too brutal for children.) It shows, in accurate historical detail, the horrible torture Jesus endured and how much He suffered for each one of us.

Luke, a physician, reports He sweats blood, a medical condition known as hematidrosis) would most likely be caused by Jesus experiencing sin (not by committing sin Himself, for He never did) as if He were the worst of all sinners and thereby cut off completely from His Father. He was taking our place, paying our price, so it stands to reason that He took on all of our sin and paid the price of sin, which is separation from God.

St. Paul supports this concept with perhaps the second most startling, incredible assertion in the Bible.[39] In his Second Letter to the Corinthians St. Paul writes:

> *For our sake, he made him to be sin who did not know sin, so that we might become the righteousness of God in him.*
>
> 2 Cor. 5:21

Think about that. Here is the most innocent, prefect human being to ever live and He is to become sin personified in order to give each one of us the chance to spend eternity in His presence in heavenly peace and joy.

This prospect of His transformation into sin itself is what I believed causes Jesus so much anxiety and fear in the Garden that night. If He becomes sin, His Father has to reject Him for sin is, in its essence, the destruction of the loving relationship that is supposed to exist between a human being and his heavenly Father. Thus, not only is Jesus to suffer horrible physical torture and the most painful of deaths; but He is going to do so utterly alone, without the comforting and reassuring presence of His Father.

The next afternoon, on the cross, when that unabated stark aloneness becomes almost too much to bear, and His suffering finally reaches a crescendo, He pulls Himself up on the nails in great pain so He can speak loudly:

> *And about three o'clock Jesus cried out in a loud voice, "Eli, Eli, lema sabachthani?" which means, "My God, my God, why have you forsaken me?"*

---

39 The most incredible assertion in the Bible is:
And the Word [Jesus] *became flesh and made his dwelling among us.*
John 1:14

Matthew, 27: 46.

This is the cry that demonstrates the total separation of the Father from His Son, Who has knowingly and willingly become our sin out of love. But, then, as He does His whole life, He immediately settles back into His sacrificial dedication to each one of us and continues to endure the cross until finally His mission is complete. At that point, He would . . .

cry out in a loud voice, "Father, into your hands I commend my spirit"; and when he had said this he breathed his last.

Luke 23:46

Thus is demonstrated the greatest of all loves – the love God has for each one of us. In the garden, Jesus knows all this physical pain and suffering is about to descend upon Him; and, much more significantly, He and the Father both know they will lose their relationship as Jesus becomes sin for our sake -- yet they both proceed with their plan for our salvation. Do you realize that by doing this for us God demonstrates that each one of us is of unimaginable and inestimable value? This is the price God is willing to pay for your soul. Oh, what a price indeed![40]

As Jesus prays and sweats blood in the garden, His disciples are totally oblivious to the drama that is about to unfold. In fact, after their large Passover meal they are very tired and fall asleep, even though Jesus twice asks them to remain awake with Him and pray.

After some hours, Jesus awakens them one final time as a large group of men come to arrest Him. At this point, His disciples -- His best friends – all desert Him and run away into the night. Jesus allows

---

40 This is the reason abortion is such a terrible sin. Jesus paid the price required to allow fully developed souls, who can know and freely choose to love God and neighbor, to enter heaven. Abortion cuts short the baby's development and he, or she, can thus never participate, to the full extent paid for by Jesus, in the loving relationship God intended for that child to enjoy with Him. Abortion has eternal deleterious consequences for the baby (and the mother, if she does not repent) and it is tragic that these realties are never mentioned in any discussion about abortion, for they last forever.

Himself to be bound up and led away. He spends the night being hit repeatedly in the face, spat upon and beat up.

In the early morning, He is taken to Pontius Pilate, the Roman procurator over Jerusalem. Israel is an occupied country and local authorities cannot impose the death penalty – only the Roman authority can do that. So, if Jesus is to die, as the local authorities desire, they need Pilate to pronounce the sentence.

Pilate interviews Jesus and finds no evidence which would justify the death sentence (or any sentence). He decides instead to "punish" Jesus, even though he has declared Him innocent, and then release Him. The idea of "punishment" in a Roman procurator's mind and the idea of punishment anyone of us might have are two very different concepts. The ancient Roman world is a cruel world where human life is cheap and of no particular value. Thus, Pilate's idea of punishment is to order Jesus to undergo the horrific penalty of scourging.

A Roman scourging is often a death sentence in itself for only the strongest of men can live through it. It is simply awful torture.

In the Roman Empire flagellation was . . . sometimes referred to as scourging. Whips with small pieces of metal or bone at the tips were commonly used. Such a device could easily cause disfigurement and serious trauma, such as ripping pieces of flesh from the body or loss of an eye. In addition to causing severe pain, the victim would approach a state of hypovolemic shock due to loss of blood.

The Romans reserved this treatment for non-citizens . . . . The poet Horace refers to the horribile flagellum (horrible whip) in his Satires. Typically, the one to be punished was stripped naked and bound to a low pillar so that he could bend over it, or chained to an upright pillar so as to be stretched out. Two lictors . . . alternated blows from the bare shoulders down the body to the soles of the feet. There was no limit to the number of blows inflicted - this was left to the lictors to decide, though they were normally not supposed to kill the victim. . . . . Flagellation was referred to as "half death" by some authors and apparently, many died shortly thereafter . . . . . In some cases the victim was turned over to allow flagellation on the chest, though this proceeded with more caution, as the possibility of inflicting a fatal blow was much greater.[41]

---

41 Wikipedia description -- http://en.wikipedia.org/wiki/
    Flagellation#Christianity

A Roman scourging left the victim with his flesh torn off his body, hanging in bloody strips. Many parts of the skeleton understructure were visible. As mentioned above, King David, in Psalm 22, written about 1,000 years earlier, describes Jesus' passion in amazing detail, including this observation:

> *I can count all my bones.*
> Psalms 22:18

That is an accurate, if somewhat cryptic, description of Jesus after the scourging. When this flesh-stripping whipping is over, a few of the Roman soldiers have a surprise for Jesus. They have heard that He is passing Himself off as the king of the Jews so they decide to give Him a crown. They fashion one out of thorny branches, jam it on His head and then, while paying Him mock homage, repeatedly hit Him on the head with a stick, driving the multitude of thorns deep into His scalp. As each of us knows, this particular torture would be incredibly painful and bloody, since our scalp is very sensitive and well-supplied with blood vessels.

Since Jesus was going to die for our sins on the cross, why did the Father's plan of salvation include the horrible tortures of scourging and crowning with thorns? Would not crucifixion be enough? Again, of course, this is a mystery; but it might be that Jesus undergoes this total disfigurement of His body and His head in order to create an outward, physical sign of the terrible inward, spiritual ravages of a soul in sin.

Just how grotesque does He appear after these tortures? When He is returned to Pilate in this bloody, flesh-torn state, even this experienced, hardened Roman leader is taken aback by the horror of the sight that greets him. In fact, He turns to the crowd, hoping to get them to back off their demand that Jesus be crucified when they view the horrific sight, and yells:

> *Behold the Man.*
> John 19:5.

But his plan fails. The crowd shows no sympathy and keeps shouting to Pilate to crucify Him. Finally, Pilate, seeing there is no hope for a change in the crowd's sentiment relents and Jesus is condemned to crucifixion.

Under Roman law, the condemned customarily carry the cross beam of their own cross to the site where they are nailed to it to die. So, Jesus starts out along the half-mile route to the place of crucifixion, outside the walls of Jerusalem, carrying His cross. As He struggles along the rough stone hewn path, He most likely kept rubbing the heavy wood beam, inadvertently, against the crown of thorns, driving thorns deeper into His skull with every step He took.

Three times He falls hard upon the ground because He is so weak due to a loss of blood and the interminable excruciating pain from the scourging and thorns. Each time the heavy wood beam must have crashed down on His head, shoulders and back sending shock waves of additional pain throughout His body. Finally, the soldiers, who had been whipping Him to get Him to keep walking forward, realize that He is simply not going to be able to physically go any further. They conscript a man from the crowd, Simon of Cyrene (a town in Libya) who helps carry Jesus' cross until they arrive at Golgotha (the place of the skull) where Roman Judean crucifixions were performed.

Once there, the process of crucifixion proceeds quickly with Roman efficiency. First Jesus' arms are outstretched and nails hammered through each wrist just above the location where the radius and ulna bones come together. This placement accomplishes two objectives: it provides a solid location which can support the body's weight without the nails tearing out and it pierces the median nerve (the nerve that causes carpel tunnel syndrome) thereby causing two additional sources of constant excruciating pain.

The cross beam is lifted up onto an upright beam planted in the ground and His feet are nailed to the upright. The left foot is pressed backward against the right foot, and with both feet extended, toes down, a nail is driven through the arch of each severing nerves running between the metatarsal bones.

Jesus now hangs between heaven and earth suspended on three pivot points of rough hewn roman nails on a wooden cross that was His destiny from time immemorial.[42] As He sinks down from the

---

42 His Father plans for Jesus' incarnation and death, on our behalf, even before He creates the universe. As Asher Intrater has written: *[T]he*

weight of His body, He suddenly realizes that He cannot exhale; his diaphragm will not work in this posture. He therefore has to pull up and push against the three nails in order to raise Himself up so he can breathe. This tormenting feature of crucifixion radically exacerbates the excruciating pain that is ever present on the cross.

The first words from Jesus on the cross are incredible in light of all the circumstances.

Then Jesus said, "Father, forgive them, they know not what they do.
Luke 23:34

Jesus is innocent of all charges; yet the first thought He has, in almost unbearable pain, is to forgive those responsible for His unjust execution. They do not know they are killing the true Son of God, the savior of us all, and therefore they will not be held accountable for this action.

For six hours Jesus hangs on the nails. He refuses to allow Himself to go unconscious or not to feel fully the pain and agony inflicted by the horrific torments and tortures to which He has been subjected since He entered the garden the evening before. This is the price that must be paid for our sins. Around noon, the sky goes dark as clearly foretold about 750 years earlier by the prophet Amos:

On that day, says the Lord GOD, I will make the sun set at midday and cover the earth with darkness in broad daylight.
Amos 8:9.

---

crucifixion of Yeshua [Jesus] was planned and determined before creation. God's love is not only perfect, it is also sacrificial. It is perfectly sacrificial. The standard that Yeshua was willing to give His life for us was set before the world began. Revelation 13:8 [states]: In the book of life of the lamb slain from the foundation of the world. Logically, the crucifixion had to happen. A loving creator makes a loving creature. A loving creature has free will. Free will allows the possibility of sin. Sin leads to death. Death can only be cured by resurrection. Resurrection can only happen to one who has died. The crucifixion and resurrection of Yeshua was preplanned before creation. See also Ephesians 1:4-6; Revelation 13:8

About 3 PM Jesus has given all He that is required to give and He dies.

When Jesus had taken the wine, he said, "It is finished." And bowing his head, he handed over the spirit.

John 19:30.[43]

Jesus is buried in a borrowed tomb. On the third day, as foretold in the Jewish Scriptures and many times by Jesus Himself during His ministry, He is resurrected from the dead. He spends 40 days on earth, is seen by hundreds of people and gives His apostles repeated instructions about carrying His Good News – that salvation is available to anyone who chooses to believe in God and to obey His two commandments to love -- to all the world after He is taken up to heaven. By His death and resurrection, our debt is paid and salvation becomes possible for each one of us.[44]

---

43 The words that a task is "finished" are written only 2 times in Scripture. The other instance is found at the very beginning of creation in Genesis. In a sense, Scripture starts and ends with the two fundamental acts of creation – the material creation and the spiritual re-creation -- both being "finished."
Since on the seventh day God was finished with the work he had been doing, he rested on the seventh day from all the work he had undertaken.

Genesis 2:2

44 Since you already have eternal life, the only question is where will you spend your eternal life? Stage one is here on earth. But your ultimate destination is yet to be determined. It is your choice for you, and only you, are responsible for the decisions and choices you make in this life. In the words of the Dalia Lama: *Human potential is the same for all. Your feeling 'I am of no value' is wrong. Absolutely wrong. You are deceiving yourself. We all have the power of thought – so what are you lacking? If you have willpower, then you can change anything. You are your own master.*

CHAPTER V

THE WORLD AFTER JESUS

When Jesus dies around 33 A.D., his small band of Apostles begin their mission to inform the Jewish nation and then, after several years, the whole world, of the Good News. Jesus came to earth and paid the price God's justice demands for all our sins and we now have the opportunity to live with Him in heaven forever, if we make loving choices in this life. Many of us still have not accepted this Good News and the opportunity it presents. This rejection is obvious for if each one of us loved God with all his heart, soul, strength and mind and loved his neighbor as himself, the world would be one of love, peace, compassion and generosity.

So, our human story since Jesus ascended to heaven has continued to be a mixed bag. There have been millions who have lived wonderful, kind, loving lives – caring for others more than themselves; sharing their time, talent and treasure with others without counting the cost. These are the saints who have lived among us and most of us are fortunate enough to know a person or two who lives this way.

Unfortunately, millions of us have chosen to walk a different path – the way of hate, bigotry, dispute, disagreement, cold-heartedness, selfishness and meanness. Jesus knew there would be these two paths continuing after His death and He knew the majority would choose the wrong one.

Enter through the narrow gate; for the gate is wide and the road broad that leads to destruction, and those who enter through it are many. How narrow the gate and constricted the road that leads to life. And those who find it are few.

St. Matthew 7:13-14.

How much He must love us to endure the torture He did to pay the price for our sins while knowing that many of us, down through the centuries, would have hardened hearts and freely choose to ignore His sacrifice. This, truly, is love beyond our understanding.

In any event, after His ascension, His message spreads rapidly around the then-known world. As the Gospel travels, fundamental theological questions arise about the nature of the Triune God, the nature of Jesus, the exact meaning of His message, the nature of evil and the like. Much time, in the first few centuries after the ascension, is spent by the leaders of the church – the ecclesia, the collective body of Christian believers -- wrestling with these weighty issues. The decisions they make are incorporated into the core beliefs of Christianity that come down to us today.

The process they devise to resolve questions that are disputatious is always the same:

- Someone, perhaps seeking a deeper personal or intellectual understanding of the faith, presents a novel or reformatted theological proposition.
- Extensive discussion ensues among believers and some support and others oppose the new statement.
- In time, it becomes apparent that various parties cannot reach agreement.
- If the contested matter impacts on a key understanding of one of the core beliefs of the Christian faith, the church leadership, usually the Pope in these early times, convenes an Ecumenical (i.e., pertaining to Church unity) Council. This is a meeting of Church leaders and theological experts from all over the then known world at which they discuss the issue and decide if the proffered proposition is an accurate statement of a basic doctrine of the faith or if it is to be declared heretical.
- The Council's decision then becomes part of the core beliefs – the dogma – of the Christian faith.

30 A.D.[45] – Fifty days after Jesus' resurrection, and 10 days after His ascension, on the Jewish (and now also Christian) Feast of Pentecost, the Holy Spirit descends upon His Apostles and other faithful followers. They are changed from fearful, inarticulate people with a limited understanding of God's plan for our salvation, into courageous, articulate and comprehending individuals who can communicate the Good News so that each listener understands it.

34 A.D. – St. Paul (formerly Saul of Tarsus – 5 B.C. to 67 A.D.) converts from being a dedicated Jewish Pharisee, and prosecutor of Christians, to a follower of Jesus. Some years later, he takes on the assignment of primary responsibility for carrying the Good News to the western world outside Israel.

50 A.D. – St. Peter has a vision that convinces him that the Good News is meant for every one in the whole world without the application of Jewish legal proscriptions to any convert. The First Council of Jerusalem is held and the Church affirms that gentiles are to be baptized without having to convert first to Judaism.

312 A.D. – Emperor Constantine declares Christianity to be official faith of the Roman Empire.

325 A.D. – The first Ecumenical Council is held in Nicaea (a city in Turkey) to resolve a hotly contested debate about the true nature of Jesus as the Son of God. The Arian heresy asserted that the Son was not co-equal with the Father. The Council declares that Jesus was consubstantial (of one and the same substance or being) and coeternal with the Father. The Nicene Creed, still recited today in Catholic and Episcopalian churches with very few revisions, is crafted and approved.

381 A.D. – The second Ecumenical Council is held at Constantinople to confront the assertion that the Holy Spirit is not truly divine. It

---

45   Another timeline of events is presented, as stated above in Chapter III, because these events lead directly to our modern world and its social structures. Your values and judgments are formed, in significant part, by the physical and social environments in which you grow up and live and we need to include a listing of the events that helped shape your environment.

finalizes the theological concept of the Trinity – God is three co-equal persons in one essence.

393 B.C. – A synod is held in Hippo (northern Africa) and, for the first time, a group of bishops lists and approves the books (i.e. "canon") that are to be included in the Bible and considered sacred scripture. Many books that are proffered for inclusion are not accepted and are now lost.

431 A.D. – The third Ecumenical Council, at Ephesus (another city in Turkey), decrees that Mary is the Mother of God. It declares the personal unity of Christ – He is not a human body with a human soul and a divine mind. He is fully human with a human intellect and human free will.

451 A.D. – The fourth Ecumenical Council is held at Chalcedon (today a district in Istanbul, Turkey) to define the two natures of Jesus. The Council declares that He is truly God and truly man.

570 A.D. -- Mohammed is born in Mecca (Saudi Arabia). He founds Islam, a monotheistic religion, which has the Qur'an (Koran) as its sacred text. The faith spreads rapidly and today Islam is the second largest religion in the world.

The Arabic word "Islam" means "submission," reflecting the religion's central tenet of submitting to the will of God. Islamic practices are defined by the Five Pillars of Islam: faith, prayer, fasting, pilgrimage and alms.

787 A.D. -- The seventh Ecumenical Council (called the Second Council of Nicaea) lays down the principles for appropriate veneration of holy images. The Council is concerned with believers giving homage, or too much credit, to icons and statutes. In response, it distinguishes between *absolute* and *relative* worship. Absolute worship is paid to any person for his own sake. Relative worship is paid to a sign or image, not at all for its own sake, but for the sake of the person or reality signified. The sign in itself is nothing, but it shares the honor of its prototype. This relative worship is to be paid to the cross, images of Christ and the saints, etc., while the worshipper's intention directs reverence and faith, in its entirety, to the reality represented.

850 A.D. – Chinese develop formula for gunpowder.

1096 – 1291 A.D. – Seven crusades are launched by Christians in an ineffective attempt to permanently re-capture the Holy Land from Muslims.

1215 A.D. – The twelfth Ecumenical Council (the Lateran IV Council) is held in Rome, Italy. It issues a final declaration against the long-running heresy (referred to as the Manichaean or Albigenses heresy) which erroneously proposed that there were two eternal co-equal principles of creation – one good and one evil. The Council also re-asserts the long-standing dogma that Jesus is truly present under the appearance of bread and wine (the technical term is "transubstantiation") in the sacrament of the Eucharist.

The Magna Carta (Great Charter) signed by England's King John and English nobility. For the first time some of the king's powers are specifically limited and shared with others.

1518 A.D. – Martin Luther writes out his complaint against improper and abusive fiscal and political practices within the Christian church and begins the Protestant Reformation.

1529 A.D. – England's King Henry VIII breaks with the Catholic Church over his divorce and founds the Church of England.

1543 A.D. -- The advent of modern science begins with the publication of two works: Nicolaus Copernicus's *De revolutionibus orbium coelestium* (*On the Revolutions of the Heavenly Spheres*) and Andreas Vesalius's *De humani corporis fabrica* (*On the Fabric of the Human body*).

1563 A.D. – Rev. William Lee invents the Stocking Frame, a mechanical device for knitting stockings and the industrial revolution begins. The ramifications are enormous. As this revolution develops and travels around the world, the majority of citizens residing in nations that take advantage of its inventions and advances, for the first time in history, are no longer required to spend long hours directly producing the necessities of life – food, clothing and shelter -- for themselves. Instead, they go to work in organized businesses, with established hours of work, and have leisure time each week to pursue their own interests. Thus, the societies in these nations become physically and psychologically divorced from the land and much less dependent on nature and her seasons for their livelihood. This independence leads many to feel less connected to God, more confident in their own powers and abilities, and less willing to simply accept religious assertions. A philosophy of mechanistic determinism (the universe is just a big machine that runs itself) becomes attractive and gains strength, reaching its apex about two centuries later with the publication of Darwin's Origin of the Species. Many come to think that God is no longer needed.

1620 A.D. – The Pilgrims, seeking religious freedom after breaking away from the Church of England, arrive in North America.

1769 A.D. -- Scotsman James Watt patents an improved version of the steam engine that supplies the mechanical power that fuels the rapid advancement of the Industrial Revolution.

1787 A.D. -- John Fitch makes the first successful trial of a 45-foot steamboat on the Delaware River; man can now travel, for the first time, independent of animal leg muscle or the forces of nature.

1788 -1791 A.D. – The United States Constitution and its Bill of Rights amendments are adopted. A new concept in governance is born – the individual has God instilled rights that the government must always respect and protect.

1796 A.D. -- Edward Jenner discovers the first vaccine when he inoculates a young boy with material from cowpox blisters to prevent smallpox.

1825 A.D. -- George Stephenson builds the first practical steam locomotive engine for railways. The locomotive provides the first advance in overland travel since horse-powered transportation was developed 6,000 years earlier. (Consider this -- George Washington could travel no faster than Pharaoh Ramses II.)

1866 A.D. – First internal combustion gasoline engine by Nikolaus

Otto makes possible personal transportation vehicles and convenient land travel over great distances.

1876 A.D. – Alexander Graham Bell invents the telephone and people begin to speak directly with each other without regard to physical proximity.

1879 A.D. – Electric light bulb by Edison removes limitations on our activities imposed by darkness.

1883 A.D. -- Standard time zones are established due to the speed of trains and the need for precise time keeping over large geographical areas. For the first time in history, people in widely separated areas go through the day using identical clock time.

1903 A.D. – The year the world becomes a much smaller place:
First instantaneous transmission from America to Europe (Marconi wireless radio);

Wright brothers invent the first controllable, heavier-than-air flying machine; and--

Trans-Pacific Ocean telephonic cable placed and President Theodore Roosevelt makes the first around the world verbal communication – it takes 12 minutes to complete the journey.

1905 and 1916 A.D. – Albert Einstein publishes his Special and General Theories of Relativity. These new insights assert the equivalency between mass and energy, posit the space/time continuum and result in a much deeper understanding of particle physics, nuclear energy and gravity. The universe becomes finite, with a beginning and an end when entropy shuts it all down a few trillion years from now.

1927 A.D. -- Philo T. Farnsworth files the patient that gives birth to television which allows people to communicate visually without regard to physical location of the parties.

1942 A.D. – Germans build the V-2, the first rocket able to reach space.

1945 A.D. – The United States builds the first atomic bomb, based on Einstein's theories.

1957 A.D. – Russia launches the first satellite, Sputnik, in earth orbit.

1960 A.D. – The Food & Drug Administration approves the first birth control pill. It is a major contributor to significant changes in family unit dynamics throughout society.

1969 A.D. – Neil Armstrong becomes the first human being to set foot on the surface of an orb that is not the earth; in this case, the moon.

1972 A.D. – The internet is born with the first public demonstration of ARPANET (Advanced Research Projects Agency Network), created by the Defense Advanced Research Projects Agency (DARPA) of the United States Department of Defense.

1972 A.D. – The Global Positioning Satellite program -- called, at the time, the NAVSTAR GPS Joint Program -- is initiated under the leadership of Bradford Parkinson, the chief architect of GPS throughout the system's conception, engineering development and implementation.

1973 A.D. -- Doctor Martin Cooper, director of research and development at Motorola, invents the technology responsible for the modern cell phone and becomes the first person to make a call on a cell phone.

1976 A.D. – Steve Wozniak and Steve Jobs form Apple Computer and create the first practical personal computer.

CHAPTER VI

WHERE DO YOU GO FROM HERE?

I shall be telling this with a sigh
Somewhere ages and ages hence:
Two roads diverged in a wood, and I --
I took the one less traveled by,
And that has made all the difference.
Robert Frost

Enter in by the narrow gate: for wide is the gate, and broad is the way, that leads to destruction, and many are they that enter in thereby. For narrow is the gate, and straitened the way, that leads unto life, and few are they that find it. St. Matthew 7:13-14

While you have been born to live forever, you will transition, at death, from life, as you now know it, to an entirely different experience of existence. But it will be you, living forever, with your own current consciousness identity, in this new reality. You will spend your life in one of two ways depending on which road you choose, with your own free will, in this life.

On the one hand, you can choose "the broad way," as so many do, and look inward – putting yourself, your own needs, wants and desires, first; disregarding the needs and hurts and requests for help from your neighbors (i.e., anyone, including strangers, in obvious need of assistance, whom you could help, if you but choose to do so). When you take this road, you form your character accordingly; and, after you die, your selfishness and pride judge you to hell.

Exactly what awaits those who choose this broader path? We have a very good picture of what awaits souls condemned to hell. In 1936, God has provided this graphic vision to Sister Faustina Kowalska, which she records in her diary:

*Today, I was led by an Angel to the chasms of hell. It is a place of great torture; how awesomely large and extensive it is! The kinds of torture I saw: the first torture that constitutes hell is the loss of God; the second is perpetual remorse of conscience; the third is that one's condition will never change; the fourth is the fire that will penetrate the soul without destroying it – a terrible suffering since it is a purely spiritual fire, lit by God's anger; the fifth torture is continual darkness and a terrible suffocating smell, and, despite the darkness, the devils and the souls of the damned see each other and all the evil, both of others and their own; the sixth torture is the constant company of Satan; the seventh torture is horrible despair, hatred of God, vile words, curses and blasphemies. These are the tortures suffered by all the damned together, but that is not the end of the sufferings. There are special tortures destined for particular souls. These are the torments of the senses. **Each soul undergoes terrible and indescribable sufferings, related to the manner in which it has sinned**. There are caverns and pits of torture, where one form of agony differs from another. I would have died at the very sight of these tortures if the omnipotence of God had not supported me. **Let the sinner know that he will be tortured throughout all eternity, in those senses which he made use of to sin. I am writing this at the command of God, so that no soul may find an excuse by saying there is no hell, or that nobody has even been there, and so no one can say what it is like.***

*I, Sister Faustina, by the order of God, have visited the abysses of hell so that I might tell souls about it and testify to its existence. . . . . I noticed one thing: that **most of the souls there are those who disbelieved that there is a hell**. When I came to, I could hardly recover from the fright. How terribly souls suffer there!*[46]

On the other hand, you can choose "the narrow gate," the road "less traveled by," as so few do, and look outward in love – subordinating yourself, your own needs, wants and desires as you try to meet the needs and requests for help from your neighbors (i.e., anyone in obvious need of assistance). When you take this road, you again form your character accordingly; and, after you die, you continue on this path, the path of

---

46 Kowalska, Sister Faustina. *Divine Mercy in my Soul.* Stockbridge: Marian Press, 1987. pp. 296-297. Print. (Emphasis added.)

love, for all eternity. You live with, and focus on, infinite Love Himself and you are completely satisfied and immeasurably happy with that contemplation; in no small part, precisely because you know it does go on without end.[47]

Sister Faustina also had a vision of heaven. She writes:

*Today I was in heaven, in spirit, and I saw its **inconceivable beauties and the happiness** that awaits us after death. I saw how all creatures give ceaseless praise and glory to God. I saw how great is happiness in God, which spreads to all creatures, making them happy; and then all the glory and praise which springs from this happiness returns to its source; and they enter into the depths of God, contemplating the inner life of God, the Father, the Son, and the Holy Spirit, whom they will never comprehend or fathom.*

*This **source of happiness is unchanging in its essence, <u>but it is always new</u>, gushing forth happiness for all creatures.** Now I understand Saint Paul, who said: "Eye has not seen, nor has ear heard, nor has it entered into the heart of man what God has prepared for those who love him."*[48]

The question for you is: on which road do you travel from this point forward? The choice is as stark as right or wrong, light or dark, love or hate, living for others or living for self. The decision you make results in you spending your eternal life living the choice of the path you decide to take now and tomorrow and the next day until you die, at which point your soul's character is formed and your choice is frozen forever.

The bad news is that, while the reward is immeasurable and eternal, the choice to love is not easy. It is very difficult to live for others, to die to the self that cries out for attention from within your very own being. We simply do not want to change our behavior and shift the focus from inward to outward, from us to others. That is the real reason we do not want to think about spiritual matters; why we do not want to study the

---

47 St. Augustine describes heaven this way. *[T]here is a sublime crated realm cleaving with such pure love to the true and truly eternal God that, though not coeternal with him, it never detaches itself from him and slips away into the changes and successiveness of theme, but rests in utterly authentic contemplation of him alone.* St. Augustine, op. cit., Kindle Edition, location 4135-56

48 Kowalska, op. cit. pp 310-311. (Emphasis added to point out why we will never be bored in heaven. God is infinite and He is always new to us so we will never finish our quest to know and love Him.)

Scriptures; why we do not want to know about, and believe in, Jesus – it means we have to change. This truth is as old as man himself.

> *And this is the verdict, that the light came into the world, but people preferred darkness to light, because their works were evil.*
> *For everyone who does wicked things hates the light and does not come toward the light, so that his works might not be exposed.*
> *But whoever lives the truth comes to the light, so that his works may be clearly seen as done in God.*

> *ST. JOHN 3:19-21*

So, is all lost if you are currently walking in darkness, on the broad road? No. There are millions of individuals, down through the centuries, who decide to change their lives and walk toward the light on the narrow path. You may even know such a person yourself. At the very least, you are most likely aware of stories about individuals who changed from being selfish to selfless and ended up spending the remaining years of their lives in loving service to others.

These are the ones who know what life is all about – who know why they were born. These "few" enter by the narrow gate and, after a few, brief years on earth, enjoy a new life filled with love and complete satisfaction forever. These are "the few" you need to emulate in the daily battle you should wage to choose right over the wrong, love of God and others over self.

If you accept the facts that

- You already have eternal life;
- There are only two possible everlasting outcomes to the choices you make now; and
- The good outcome is not easy, but requires hard work to achieve it --

then you have to ask yourself if you are able to do it.

The answer is a resounding YES. But, as is true for any goal worth achieving, it cannot be done on autopilot and you need a strategy. Think about the individuals who live a life of loving, selfless choices for others. It is almost a certainty, if you have knowledge of their private lives, that you know that it is not easy for them to live such generous lives. You have to wonder how they able to live with love for others that is equal to the love they have for themselves.

While each such loving individual undoubtedly has his or her own motivations, the one characteristic most common to virtually all such persons is faith. These loving people believe in God, believe that God loves them, and believe that God commands that they love Him and their neighbors. They then try, every day, to live by these commandments. Faith is the key.

But what is faith? Faith is the reasoned acceptance and affirmation of the transcendent reality of God. While faith does involve a leap to accepting truth that cannot be empirically verified by scientific method, it is also, and essentially, a reasoned acceptance. My reasoning for my faith has been stated throughout this book. In summary, it is this.

I. I believe in God Who is a Person

1. The universe exists, although it does not have to; and, in fact, there was a time when it did not exist. It is thus contingent being -- being which may, or may not, exist.

2. Contingent being is always the created effect of an efficient cause, and this includes the universe itself.

3. This efficient cause cannot be contingent or else it would be caused by another cause and so on ad infinitum, and not ultimately explain THE FACT that the universe exists. Therefore, the ultimate efficient cause must be an uncaused cause; i.e., a cause which has as its essence existence, "beingness." It simply (and incomprehensively) IS.

4. You and I are also continent beings. Our existence is caused by an efficient cause. The proximate efficient cause is our parents; the ultimate efficient cause is the uncaused cause.

5. You and I have intelligence and free will. This means each one of us is a person. Our minds can abstract to know good and evil and our free will gives us the power to choose good or evil. We are aware of these powers and accept them as true. (In fact, our entire criminal law system is premised on the existence of these faculties in each one of us.)

6. Since we are effects of an ultimate efficient uncaused cause, and since every effect bears the imprint of its cause, our ultimate uncaused cause also has intellect and free will.

7. This ultimate efficient uncaused cause is without limit in being and without limit in intellect and free will. He is a Person, infinite in all respects. This infinite, uncaused cause is Who I call God.

II. I believe in my soul and eternal life

1. Our faculties of intelligence and free will can transcend the material world and enable us to soar into the realm of the non-material, the purely spiritual. We can abstract to the third degree and contemplate purely spiritual concepts such as God, love, generosity and the like.
2. There is a spiritual component to our being in which adheres these spiritual faculties. This component is called our soul.
3. Since our soul is able to engage in purely spiritual activities and it is not ultimately dependent on materiality to carry on its activities, it is designed to live beyond its matter-related life here on earth.
4. When we leave this material world, our soul lives on with our own identity, performing the spiritual activities of knowing and choosing, forever.

III. I believe in Jesus Christ and the Trinity

1. God, Who is a person and Who created the universe and us, has communicated to us down through the centuries. This is what persons do – they communicate one with another.
2. This communication began, in its fullness, some 3,850 years ago when God picked Abraham to found the Jewish national identity. This group was established to be vehicle through which He would work His plan of our salvation for He loves us and cares about us.
3. For more than 1,800 years, God interacted with various Jewish leaders and prophets – e.g., Moses, David, Isaiah, Jeremiah, Ezekiel etc. – refining this group's faith and understanding of His nature and His plan.
4. When the time was right, God sent His Son, Jesus, to become true man while remaining true God, to accomplish two objectives:
    a. To reveal to us a much more detailed knowledge of the nature of God -- that God is actually three Persons in one Being Who is pure love -- and of our own ultimate destination; and, more importantly,
    b. To suffer and die in His humanity, and be brought back to life on the third day, in order to satisfy God's

attribute of justice and thereby allow us to participate, once again, in eternal life with God, if we obey His commandments of love.

5. Jesus is the final, formal direct communication with us since in Him is revealed all we can know about God in this earthly life. His communication has been memorialized in the Sacred Scriptures -- the New Testament of the Bible, as established and interpreted by His Church.

6. The bottom line message of Jesus is this:
   a. God is love and He loves each one of us with a love that is beyond our capacity to know and understand;
   b. We are to respond to His love by loving Him with our whole being and by loving our neighbor – anyone who needs our help – as He loves us;
   c. The more we live a life of loving God and our neighbor, the more we become like Him, for we slowly transform ourselves into a loving being too;
   d. If we become persons of love, our souls are then welcomed in His Heavenly Kingdom when we transition from our earthly life to our eternal destination. If we fail to love God and our neighbor, then we slowly transform ourselves into selfish beings who are only interested in ourselves and that is how we will spend eternity – lonely, focused on the self and in agonized suffering as we contemplate, forever, what could have been if we had only exercised our self-discipline and denied ourselves while reaching out to God and our neighbors in love.

This then is the meaning of life and why you were born. Please take the time and make the effort to develop a life of faith. Pray to God for a strong faith in Him Who loves you more than you can even imagine. Respond to that love by loving Him and serving your neighbor.

Believe that God is merciful. When you sin – sin is any action or thought that disrupts or weakens your relationship with God – repent, i.e., be truly sorry and committed to trying your best to not do so again. We are all sinners. Some sin and ask God for forgiveness in true sorrow. Others sin and do not ask for forgiveness either because they underestimate God's love and mercy or because they turn inward and refuse to recognize the need to repent. Sin without repentance leads to eternal misery. Sin with true repentance leads to everlasting happiness.

One final note needs to be stated. You have read this short book. Please take its content seriously. You now know more than you did before you read it. With increased knowledge comes increased responsibility and liability. Jesus put it thus:

I came into the world as light, so that everyone who believes in me might not remain in darkness.

And if anyone hears my words and does not observe them, I do not condemn him, for I did not come to condemn the world but to save the world.

Whoever rejects me and does not accept my words has something to judge him: the word that I spoke, it will condemn him on the last day,

because I did not speak on my own, but the Father who sent me commanded me what to say and speak.

And I know that his commandment is eternal life. So what I say, I say as the Father told me.

ST. JOHN 12:46-50

If I had not come and spoken to them, they would have no sin; but as it is, they have no excuse for their sin.

Whoever hates me also hates my Father.

ST. JOHN 15:22-23[49]

Jesus has revealed to us that the God-Creator of the universe, this wondrous place we discussed at the beginning of this book, is, much more importantly, a God of love. He has created you with eternal life in the hope that you will use your free will to become an individual who is also a person of love. In this way, you become eligible to spend the rest of your life – that is, all eternity – in His presence, bathed in His love, contemplating His infinite goodness and in perfect internal and external harmony and completeness. What a life awaits those of us who learn to love Our Father and our neighbor!

---

49 Bishop Sheen explained this reality thus: *As scientific truths put us in an intelligent relation with the cosmos, as historic truth puts us in temporal relation with the rise and fall of civilizations, so does **Christ put us in intelligent relation with God the Father;** for He is the only possible Word by which God can address Himself to a world of sinners.* Sheen, op. cit., Kindle Edition, location 3582-87. (Emphasis added.)

EPILOGUE

The important thing is not to stop questioning. One cannot help but be in awe when he contemplates the mysteries of eternity, of life, of the marvelous structure of reality. It is enough if one tries merely to comprehend a little of this mystery every day. Albert Einstein

We start with a quote from Einstein and we end with one. His advice is as sound in this philosophical statement as his formulas were ground braking in his mathematical texts. You have an intellect. Spend some time each day thinking about the important issues, those with eternal consequences. Use your mind to investigate the amazing world around you. See if you do not find God's footprints everywhere you look. Watch a flower bloom, a bird fly, a porpoise swim, a cloud form, another human being do a gratuitous act of kindness, and ask: what, or who, is the origin of these marvels?

I wish you the best as you continue on your journey in this world and the next. I hope you find answers that satisfy your sense of logic, common sense and wonder. We live in exciting times and data is being generated every day that demonstrate just how marvelous and wondrous is our world and us human beings too! As you search and meditate on these basic life questions, you may wish to say this prayer from time to time:

Dear Jesus: You tell us -- ask and it shall be given to you; ask and you shall receive. Therefore, believing in this promise:

  o *I ask, dear Jesus, with all the faith I possess, while beseeching You to increase my faith each day, and to make up what is lacking in my faith today with your grace so that I ask in perfect faith:*

   That I know You and that I know you better each day;

That I believe in You and that my belief in You grow each day; and

> *That my knowledge of, and my belief in, You consume my soul and You be the focus of my entire life.*

- ○ *I ask, dear Jesus, with all the faith I possess, while beseeching You to increase my faith each day, and to make up what is lacking in my faith today with your grace so that I ask in perfect faith:*
  - ○ *That I know and believe, really believe:*
    - *That You love me with a love that surpasses all human knowledge, understanding and wisdom;*
    - *That You love me with a love that is so profound and so deep that I cannot even begin to imagine such love;*
    - *That I respond to Your great love for me, and to the countless blessings and graces You continuously pour out upon me, by loving You with all my heart and with all my soul and with all my strength and with all my mind, and by loving my neighbor as myself.*
- ○ *Finally, I ask, dear Jesus, with all the faith I possess, while beseeching You to increase my faith each day, and to make up what is lacking in my faith today with your grace so that I ask in perfect faith, when the time comes for me to stand before Your Throne and be judged by the life I have lived on earth:*
  - ○ *That you have mercy on my soul;*
  - ○ *That you forgive me all my sins and cast them as far from me as the east is from the west; and*
  - ○ *That You welcome me home to heaven to live with You, Our Father and the Holy Spirit, face-to-face, forever and ever. Amen.*

APPENDIX A

Just how much precision is enough to convince our scientists that the only logical explanation for our universe is an intelligent designer/creator?

Our universe is only possible due to a balance among various forces that is so precise even the slightest deviation would have rendered it essentially different from the way it is.

Examples of this incredible precision are set forth in the book -- *Just Six Numbers: The Deep Forces That Shape the Universe*, by Martin John Rees, an English cosmologist and astrophysicist. What are his six cosmological numbers that require amazing precision for our universe to exist?

1. When hydrogen converts to helium it must do so in a process that coverts seven one-thousandths of its mass to energy. This number represents the strength with which atomic nuclei bind together and it is how all atoms were made. Just one one-thousandths variation either way, and bye-bye our universe. For example, if the conversion process only converted 0.006 percent of the mass, no transformation would occur and the whole universe would consist of hydrogen, period. Conversely, if 0.008 percent were converted, hydrogen would have ceased to exist (one consequence of this, no water anywhere).

2. The electrical forces that hold atoms together, divided by the force of gravity, must be a precise strength. If the strength of these forces were only a little smaller, there would have been created a short-lived, very small universe; and life, if it were created, would be limited to insect-size organisms.

3. Antigravity is theorized to set the limits for the expansion of our universe. It acts on a very large scale and is, itself, a very

weak force. But its role is crucial; and, again one of great precision. If it were any more potent, the present universe could not exist.

4. The cosmic number equals the amount of material in the universe. If it were a little higher or lower, the universe would either have collapsed long ago or would contain no stars or galaxies.

5. The ratio of two fundamental energies has a value is about 1/100,000. If it were a little less, the universe would be inert and with no structure; if it were a little larger, vast black holes would rule the day at the expense of all our stars and galaxies.

6. Finally, the number of spatial dimensions in our universe is 3 and that number happens to be precisely correct for life – there is no life if there were either 2 or 4 such dimensions.

All of this precision cries out for the logical conclusion that an infinite intelligent designer must be the creator of this universe. To say that we just lucked out and the universe just somehow came together by chance with all this precision and order defies common sense and logic much more than does the assertion that such a designer is responsible for our universe. Which makes more sense to you?

APPENDIX B

ODDS AGAINST ONE SPECIES EVOLVING INTO ANOTHER --
1 FOLLOWED BY 2,700 ZEROS

1/100, 000,

Odds against winning the Florida Lottery -- 1/13, followed by 6 zeros

1/13, 000, 000,